Franklin Hiram King

The Soil, its Nature, Relations, and Fundamental Principles of Management

Franklin Hiram King

The Soil, its Nature, Relations, and Fundamental Principles of Management

ISBN/EAN: 9783744757614

Printed in Europe, USA, Canada, Australia, Japan

Cover: Foto ©Suzi / pixelio.de

More available books at **www.hansebooks.com**

THE SOIL

ITS NATURE, RELATIONS, AND FUNDAMENTAL
PRINCIPLES OF MANAGEMENT

BY

F. H. KING

PROFESSOR OF AGRICULTURAL PHYSICS IN THE
UNIVERSITY OF WISCONSIN

New York
MACMILLAN AND CO.
AND LONDON
1895

EDITOR'S PREFACE TO THE RURAL SCIENCE SERIES.

The rural industries have taken on a new and quickened life in consequence of the recent teachings and applications of science. Agriculture is no longer a mere empiricism, not a congeries of detached experiences, but it rests upon an irrevocable foundation of laws. These fundamental laws or principles are numerous and often abstruse, and they are interwoven into a most complex fabric; but we are now able to understand their general purport, and we can often trace precisely the course of certain minor principles in problems which, a few years ago, seemed to be hopelessly obscure, and which, perhaps, were considered to lie outside the sphere of investigation. Agriculture has developed into a system of clear and correct thinking; and inasmuch as every man's habit of thought is determined greatly by the accuracy of his knowledge, it follows that the successful prosecution of rural pursuits is largely a subjective matter. It is therefore fundamentally important that every rural occupation should be contemplated from the point of view of its underlying reasons. It should be approached in a philosophic spirit. There was an attempt in the older agricultural literature to discuss rural matters fundamentally; but the knowledge of the time was insufficient, and such writings fell into disrepute as being unpractical and theoretical. The revolt from this type of writing has given us the present rural literature, which deals mostly with the object, and which is too often wooden in its style. The time must certainly be at hand when the new teaching of agriculture can be put into books.

For many years the writer has conceived of an authoritative series of readable monographs, which shall treat every rural problem in the light of the undying principles and concepts upon which it rests. It is fit that such a series should be introduced by a discussion of the soil, from which everything ultimately derives its being. This initial volume is also an admirable illustration of the method of science, for the soil is no longer conceived to be an inert mixture, presenting only chemical and simple physical problems, but it is a scene of life, and its physical attributes are so complex that no amount of mere empirical or objective treatment can ever elucidate them. If the venture should prove that the opening century is ready for the unrestrained application of science to rural life, then it is hoped that the Rural Science Series, under the present direction or another's, may ultimately cover the whole field of agriculture.

L. H. BAILEY.

CORNELL UNIVERSITY,
   ITHACA, N.Y., June 1, 1895.

PREFACE.

In the preparation of the pages which follow, the writer has endeavored to have them bear to the reader a rational presentation of the fundamental principles of the soil as they relate to the immediately practical aspects of agriculture. The technicalities of the subject matter, and the lines of experimentation which have contributed the facts used, have been largely avoided, not because they are deemed unimportant, but in the hope that by so doing there might result a thirst for wider reading which would lead to a search for these matters in places where they are better presented than they could be here.

No effort has been made to treat subjects in an exhaustive manner, the aim being simply to use so much of recorded facts as shall sufficiently enforce those principles underlying the management of soils which it is needful to understand in order that a rational practice may follow. The soil has been considered as a scene of life, where altered sunshine maintains an endless cycle of changes, rather than as a mere chemical and mechanical mixture, and so far as possible the problems have been given definiteness by treating them quantitively.

A free use has been made of all available literature, and credit is usually given by author's name in the text where the reference is made.

Special acknowledgment is due to the United States Geological Survey for the use of cuts in Chapter I., and also to the National Geographic Society for Fig. 11.

F. H. KING.

UNIVERSITY OF WISCONSIN,
    MADISON, WIS., May, 1895.

CONTENTS.

INTRODUCTION.

                                                                 PAGE

SUNSHINE AND ITS WORK. Nature of sunshine — Absorption and transformation of sunshine — Sunshine the motive power in winds, in evaporation, and in plant growth — The work of sunshine expressed in horse power — Movement of water in the soil and the circulation of sap influenced by sunshine — The complex character of sunshine . 3

THE ATMOSPHERE AND ITS WORK. The weight of the atmosphere — Part played in physiologic processes — Depth and density of the atmosphere — Its composition — Influence of the atmosphere on the mean temperature of the earth — Selective power of the constituents of the atmosphere — The distribution of water and plant food through wind currents 9

WATER AND ITS WORK. Part played in cooling the earth — Influence of tides upon the rate of rotation of the earth — Its work in corroding, dissolving, and transporting the materials of land areas — The part of water in the physiologic processes of plant and animal life 15

LIVING FORMS AND THEIR WORK. The protective and destructive effects of life on land areas — Part life has played in rock building and in producing mineral deposits — The great number and variety of life forms immediately important to agriculture 18

OVER AND OVER AGAIN. Cycles in nature — Conservation of energy — The circulation of the atmosphere — The magnitude and extent of water movement — Rotation of the materials of land areas 21

CHAPTER I.

THE NATURE, FUNCTIONS, ORIGIN, AND WASTING OF SOILS.

Nature and general composition of soil — Soil and subsoil — Difference between subsoils of humid and arid regions — Effect of lime on arid soils — The functions of soil — Plants without true roots — Influence of soil in the processes of evolution — The soil a water reservoir — The soil a laboratory — Origin of soils — Agencies in the formation of soil — Transition from rocks to soil — Methods of rock disintegration — The action of streams in soil growth — Sediments moved by streams — The shifting of water courses — Mechanical action of rains — Bad lands of Mississippi — Overplacement of soils — Glacial action in soil production — Part played by animals in soil formation — Formation of humus — Origin of swamp soils — Wind-formed soils 27

CHAPTER II.

TEXTURE, COMPOSITION, AND KINDS OF SOIL.

Mechanical analyses of soils — Importance of the size of soil grains in land values — Surface area of a cubic foot of soil — Influence of size of soil grains on the rate of solution of plant food — Influence of texture on drainage and aëration — Chemical elements in soils — The composition of soils as shown by chemical analyses — Sandy soils compared with clay soils — The interpretation of chemical analyses of soils — Soils and subsoils contrasted — Soils of humid and arid regions compared — Chemical composition and functions of humus — The nitrogen content of humus in arid and humid soils — The functions of certain chemical ingredients of soil — Kinds of soils — Relation of plants to different types of soils — Relation between plant food stored in the soil and that removed by crops — The "running out" of soils 70

CHAPTER III.

NITROGEN OF THE SOIL.

PAGE

Nitrogen content of soils of different regions — Nitrification in humus — Great importance of nitrogen, sulfur, and phosphorus in plant life — The quantitive relation of the nitrogen in crops compared with their ash ingredients, and these with the natural supplies in the soil — Forms in which nitrogen occurs in the soil — Distribution of nitrogen in the soil — Distribution of nitrates in the soil — Sources of soil nitrogen — Nitrogen compounds derived from the air — Sulfuric acid derived from the air compared with crop demands — Absorption of ammonia from the air by soils — Free-nitrogen-fixing germs — Different varieties or species of germs — Symbiosis — Observations of Frank, Schlösing, Jr., Laurent, and Kosswitsch on soil algæ — Processes of nitrification — Conditions which bring about denitrification 107

CHAPTER IV.

CAPILLARITY, SOLUTION, DIFFUSION, AND OSMOSIS.

Illustrations of capillarity — Nature of capillarity and relation to surface tension — Strength of surface tension — Rise of liquids in capillary tubes — Capillary movement of water in soils — Measure of capillary work — Nature of solution — Solution of plant food from soil grains — Source of power in solution — Examples of osmosis — Measurement of osmotic pressure — Method of osmotic movement — Translocation of starch, etc., in plants — The so-called selective power of plants 135

CHAPTER V.

SOIL WATER.

Functions of soil water — Amounts of water demanded by crops — Rainfall in most countries insufficient for the largest

PAGE

yields — Capacity of the soil for water — Rate of percolation from soils — Amount of soil water available to plants — The water table — Relation of water table to the surface — Wells — Contamination of well water by percolation — Movements of soil water — Rate of percolation through different kinds of soil — Capillary movements of soil water in field soils — Influence of fertilizers on the rate of capillary movement — Barometric oscillations of the ground water — Temperature oscillations of the ground water . . . 154

CHAPTER VI.

CONSERVATION OF SOIL MOISTURE.

Amount of water retained by field soils — Influence of plowing land on the loss of soil water — Importance of early tillage — Importance of early seeding — Use of catch crops to diminish the loss of fertility — Danger in the use of catch crops — Comparative losses of water on land cultivated and land not cultivated — Comparison between soil mulches of different depths — Mulching effect of soils of different kinds — Translocation of soil moisture caused by rains — Effects of cultivation on translocation of soil moisture — Translocation produced by firming the soil — Deep tillage to conserve soil moisture — More water available on drained lands — Flat and ridge culture — Influence of wind breaks and grass lands on the rate of evaporation . . 184

CHAPTER VII.

DISTRIBUTION OF ROOTS IN THE SOIL.

Distribution of corn roots under field conditions — Vertical distribution of roots — Extent of root foraging in soils — Influence of soil texture on symmetry of development . 207

CHAPTER VIII.

SOIL TEMPERATURE.

Importance of right soil temperature — Temperature at which vegetation becomes active — Observed soil temperatures at different depths — Influence of temperature on solution and diffusion in the soil — Influence of temperature on soil ventilation and upon osmosis — Temperatures favorable to germination — Influence of soil temperature on nitrification — Conditions which influence soil temperature — Specific heat of water and of soils — Influence of evaporation on soil temperature — Temperatures on drained and undrained soils — Variation of temperature with kinds of soil — Variation of temperature with direction and amount of slope — Influence of color on soil temperature — Influence of rolling and smoothing on soil temperatures — Temperature modified by depth of cultivation — Percolation after rains modifies soil temperature — Means of hastening the rise of soil temperatures early in the season — Thorough tillage — Rolling — Tillage to diminish the diurnal range of soil temperature 218

CHAPTER IX.

RELATION OF AIR TO SOIL.

Need of air in soil — Floating gardens and water culture — Saltpetre farming — Aëration of soil to prevent denitrification — Warington's experiments on water-logged soil — Soil ventilation to admit free nitrogen — Natural processes of soil ventilation — Blowing wells — Large ventilation of certain types of soils — Means of controlling soil ventilation — Flocculation increases soil ventilation — Influence of underdraining on soil ventilation — Aërating power of clover and other forms of vegetation — Soils may be too thoroughly ventilated — Hygroscopic moisture of soils . 239

CHAPTER X.

FARM DRAINAGE.

Area of swamp lands in the United States — Reclamation of swamp lands in Holland — Lands which may be improved by draining — Drainage systems in Illinois — Necessity of draining water-logged soils — Depth to which the water table should be lowered — Provision to prevent too rapid loss of water through tile drains — Distance between tile drains — The gradient of the ground-water surface — Rate of lowering the level of the water table — Natural sub-irrigation — Value of such lands for intensive farming — Needs and methods of surface drainage — Celtic land beds — Draining basins without outlets — Example of field underdraining — Fall of drains — Cost of tile draining — Size of tile — Outlet of drains — Joining laterals with mains — Obstruction of tile drains by the roots of trees . . . 253

CHAPTER XI.

IRRIGATION.

Encouragement for irrigating in humid climates — Results of sewage irrigation — Effect of irrigation on yield of corn in Wisconsin — Irrigation of grass lands in Europe — Development of irrigation in early times — Extent of irrigation at the present time — Amount of water used in irrigation — Methods of obtaining water for irrigation — Cost of irrigation — Irrigation of barren sands in Belgium . . . 268

CHAPTER XII.

PHYSICAL EFFECTS OF TILLAGE AND FERTILIZERS.

Importance of good tilth — Effect of cultivation on the texture of the soil — Seeding to grass tends to restore texture similar to that of virgin soil — Change of texture by puddling

— The formation of clods — How plowing affects tilth — Early tillage after rains to preserve good tilth — Subsoiling in humid and semiarid regions — Winter weathering to improve tilth — Burning and paring — Influence of fertilizers of different kinds in altering the texture of soils — Capillary power and rate of percolation affected by fertilizers — Physical action of soils in retaining certain salts — Influence of farmyard manure on soil moisture — Influence of summer fallowing on the relation of soil to water — The Lois-Weedon system of tillage — Summer fallowing in humid and semihumid climates 276

THE SOIL.

INTRODUCTION.

It was early one morning late in October after there had been several very severe frosts that a fox squirrel, either by chance or in deliberate search, passed under a large tree and found the ground thickly strewn with butternuts. All night these nuts had been falling by ones and by twos until now the ground was nearly covered with them. As some other squirrel had done, one, or maybe, two hundred years before, so did this one take a nut, and hurrying off to a secluded spot, bury it in the soil beneath the forest mold. Why this was done, whether with the intention of recovering it for a future meal, or whether, like a deliberate forester, he planted it that another tree might grow, only that squirrel knew. It lay there in the ground undisturbed the winter through; but in the spring, as with a thousand seeds of other kinds, its obstinate shell opened without a jar or sound. Water crept in, and the rich oil stored all winter in the thick meat rapidly changed into sugar, so that out of this and other materials borne along in streams of water which now were setting in from the soil, the tiniest cells began to form, some building a stem upward into the sunshine and air, and others building rootlets downward and outward into the darkness and dampness of the soil. As the building, or growth, went rapidly on, it was not long before the materials stored in the nut and which induced the squirrel to carry it away, had all been used; but not, however, until, as a result of this use, there stood in the rich, dark mold a perfect little butternut tree with its roots brought into contact with the water among the soil grains and its green leaves spread where the throbbing pulses from the rising sun shall be made to pump the water and do the work of building a great forest tree.

The processes of sprouting, budding, growing, and fruiting are marvelous ones, and the farmers, gardeners, and florists whose lots are cast among them have the grandest of opportunities for enjoyment and for intellectual and moral uplifting, as well as for pecuniary profit, if only they will train themselves to take advantage of them and then allow themselves the opportunity of doing so.

Believing fully in the soundness of these words, the writer, in preparing this treatise, has aimed to present the most practical and fundamental facts and principles concerning the soil as largely as possible from the standpoint of the How? and the Why? and at the same time to pause now and then to view some of the wonderful adaptations of structure to physical environment which the long processes of evolution have finally produced. We can well afford to do this because the future development of agriculture can be made most rapid and most sure, not more by giving to the farmer new facts than by making him able to observe, interpret, and correlate the facts which each and every year's planting, hoeing, and harvesting must inevitably bring under his own eyes. The business of farming has now become so complex, the sciences to which it must look for direction are so numerous, and the needs of the world for great quantities of materials for cheap, wholesome food and clothing are growing so rapidly more urgent that the farmer of Nineteen Hundred must rise upon a plane of better directed efforts and more economic methods. He can no longer do as most of us say the squirrel did, — plant without thought of adaptation or fitness, but simply as and because his father, grandfather, and great-grandfather did.

SUNSHINE AND ITS WORK.

Let us not drop at once into the soil and lose ourselves in the darkness of its details, but first let us look about and see how our field is related to the world at large and to the powers that energize in it. Let us begin with sunshine and the work it does.

While we are yet a long way from fully comprehending the nature of sunshine, we have learned enough to know that it is a sort of motion which comes to us from the sun, traveling through interplanetary space at the rate of 186,000 miles in a second of time, and on reaching us and being transformed into one or another form of energy, it does almost the entire work of the world. When the skillful man strikes a ball with a bat, he does it in such a manner that almost the entire motion of the bat passes into the ball, sending it far into the field, while the bat is brought as completely to rest at his side as if it had struck a bank of plastic clay. Now the sunshine, coming to us from the sun, is as definitely a source of power and is as capable of doing work, of setting something in motion, as is the bat when actuated by the powerful arms of the man. When the solar energy, or sunshine, falls upon the soil, the soil grains take up or absorb a portion of it as definitely as does the ball absorb the motion of the bat, the molecules making up the grains of soil have their velocities increased just as the ball had its motion as a whole augmented, and this increase of molecular motion in the soil grains is what raises its temperature.

When the motion in the surface grains becomes so great that the soil is warm or hot, some of that motion is transmitted to the molecules of air coming in contact with the earth, and these, traveling faster than they did before, push one another farther apart, thus causing the air to expand and so become lighter, bulk for bulk, than the air which has not been so heated. Once in this condition, the lighter air is forced upward by that which is heavier, and wind is the result. In this wind we unfurl the sails of a vessel or set a windmill, when the motion of the wind becomes transformed into a motion of the mill. This mill attached to a pump drives the piston, and water is raised from the well. Thus it is the sunshine warms the soil, the soil expands the air, the air drives the mill, and the mill lifts the water.

Take another case: When sunshine falls upon water, its surface molecules are set into such violent motion that the force of cohesion is overcome and the water changes to a gas, or evaporates. These rapidly moving molecules of water vapor rise quickly into the upper regions of the air where, as their motion slows down, the force of cohesion gains the ascendency again and they coalesce, forming clouds and finally large drops of rain, or flakes of snow, and fall back to the ocean again or upon the land, giving rise to soil moisture, to springs, rivulets, and rivers, or, where the temperature is permanently low enough, to ice-fields and glaciers, all of which, as we shall see, have had their part to play in the production of soil.

But not all the sunshine goes directly back after being transformed at the surface. Some of it, in its altered guise, spreads downward in the soil, and after the snow is gone and while spring is advancing into summer and summer into autumn, a large amount of sunshine is being stored. It is stored by increasing the rate of motion in the soil, and in the lower atmosphere by increasing its temperature. Now, in the temperate and polar zones, when the days become shorter than the nights, the rate of molecular motion so much slows down, because more altered sunshine is lost during the night than is received during the day, from both the surface air and surface soil in which plants live, that finally the intensity of molecular swing becomes too feeble to maintain longer the processes of growth, and plants begin to ripen, to shed their leaves, and finally, as the motion in the medium in which they live becomes too feeble, they fall asleep and winter has come. Conversely, too, when the days become longer than the nights, when more blows of sunshine reach the soil during the light than can escape during the darkness, there finally becomes an amount of molecular hustle and bustle in which sleep is no longer possible, and spring, with all its fresh verdure and joyous music, bursts upon us.

It is very essential that we fully grasp the important part which both direct and altered sunshine play in plant growth; for while we have no control over the amount which may come to a given field, we can and do control the amount stored in the soil, and we also determine the number of plants which shall stand upon the field to utilize the sunshine which does come to it.

In constructing a house, it is not sufficient to have upon the ground stone, sand, lime, lumber, and nails, enough and to spare, together with the master builder who knows how these should be put together. In addition there must be a moving power, a source of energy, which is capable of raising these inert materials and bearing them to their places. So in building the butternut tree, it was not enough that the squirrel should plant the nut in a fertile soil with abundant moisture; nor yet that there lived in the midst of that oily meat a master butternut builder capable of directing how the carbon, oxygen, hydrogen, nitrogen, phosphorus, and other materials should be compounded and brought together so that by no possible slip an acorn could drop from the boughs of the spreading tree; but, in addition, there must be a moving power, a source of energy, and this is the ether waves, born in the limitless ether ocean at the fiery surface of the sun so rapidly that more than 400 million millions of them arrive at each leaf every second, having come across 93 millions of miles in about eight minutes of time. It is under such hurried strokes as these that starch and sugar are made in the cells of plants and that cellulose is set in the framework of the tree.

So powerful is the work of the sun against the cold ether of space that Lord Kelvin estimates it at 133 thousand horse power for each square meter at the sun's surface, and the working capacity of a cubic mile of sunshine near the surface of the earth he places at nearly twenty-two horse power. Now, since the cubic miles of sunshine are arriving at the earth at the rate of 186,680 per second, one section of land and the air resting upon it must receive

$$186,680 \times 12,050 = 2,249,494,000 \text{ ft. lbs.},$$

and this is equivalent to about one-seventh of a horse power for each square foot of surface. Not all of this power reaches the earth's surface, nor is all which strikes the surface converted into work. It is not the chemical work in the cells simply which is actuated by the sunshine, but, as we shall see in another place, both the circulation of sap and the capillary flow of water in the soil are dependent upon it, in part, at least.

There is another fact regarding sunshine which we need to understand. It is this: The waves which come to us from the sun are very complex in their character; indeed, it appears as if there were a very large number of them differing from one another chiefly in their length or, what amounts to the same thing, in the number which arrive in a given interval of time, much as we know to be the fact with musical tones differing in pitch. In the case of light there is a long series of wave lengths which are characterized by being able, when they fall upon the retina of the human eye, to produce the physiological effect which we designate as color of one sort or another. But associated with these color waves, there are many others to which the human eye is not sensitive, and these are designated as dark waves. Some of these are much shorter than the color waves, and are specially powerful in breaking down the molecular structure of many substances; that is, in producing chemical changes. Then again, on the other side of the light waves, there are dark ones of much longer periods of vibration, and these long waves, to which we are insensible through the sense of sight, have a wonderful power of heating many substances when they fall upon them, and one of these substances in which we are specially interested in this study of soils, is water. When you take a large lens and let the bright sunshine pass through it, the glass is very little warmed by the passage, but if you hold paper at the light focus, it is quickly set on fire by the strong heating power of the dark or invisible rays. This has been proved by allowing the light to pass first through a solution of iodine in bisulphide of carbon, which is very transparent to the dark waves, but opaque to the light ones. When such waves are brought to a focus, they produce intense heating effects, and water is made to boil quickly under their influence. If, on the other hand, the light from the sun is first passed through a solution of alum in water, which is largely opaque to the dark waves, these are sifted out, and when the light waves by themselves are focused upon the water or paper, very slight heating effects are observed.

The fact that water is opaque to the dark rays from the sun,— that is, absorbs them instead of transmitting or reflecting them unaltered,— lies at the foundation of the evaporation of it from ocean, lakes, and streams, and also from the soil and the leaves of vegetation in some degree. When these waves fall upon the water, they set its surface molecules in rapid vibration, that is, they heat them, and this increased speed of to-and-fro movement overcomes the force of cohesion and the molecules fly off, or evaporate, as we say. Were the water not opaque to these dark waves from the sun, neither snow nor ice would be rapidly melted in the spring, nor would there be as much evaporation from the ocean as we now have, and hence rains would be less frequent and the lands less productive.

THE ATMOSPHERE AND ITS WORK.

We have dwelt very briefly upon the nature of sunshine and the part it plays in the work of the world. Let us next look at the atmosphere, for this, both in a broad way and in many details, has to do with our subject, the soil.

In the first place let us note that, light and imponderable as the air seems to be, it nevertheless presses very heavily upon all surfaces at or near sea level — so heavily, indeed, that the pressure is nearly 15 pounds on each square inch of surface and more than a ton to the square foot. Each square rod of land lying near sea level sustains a load amounting to 289 tons, and each acre 46,200 tons of air. Does it seem to you strange that, while the extended palm of one hand is pressed downward by a load of air equal in weight to that of the body, we are yet unconscious of it? If this puzzles you, place your hand beneath the surface of a pail of water. The water does not seem to bear it down. Carry your hand to the bottom of the pail. Still you do not realize that it presses any harder, and yet the bottom of the pail is carrying a load of some 20 pounds, and this is evident enough to you when you hold the pail upon the hand. When the hand is immersed in the water, it is buoyed up with a pressure equal to that which bears it down, and the pressure which crowds it to the left equals that which would move it to the right and, so long as the pressures are equal in all directions, we are unconscious of them. As fish move through the waters of the stream, the lake, or the ocean, unconscious of and unimpeded by the pressure of it, so do we travel along the bottom of an ocean of air; for such in reality the atmosphere is.

In breathing, in drinking, and also in eating, we regularly, but unconsciously, utilize atmospheric pressure. Raising the ribs and lowering the diaphragm lifts off the outside air-pressure in part, and at once the air crowds in until the lungs are stretched and filled by it. To drink, we project the lips under water, shut off communication with the nostrils, using for this purpose the curtain called the soft palate, and then draw down the floor of the mouth, thus making the cavity larger; this removes part of the pressure from the water inside the lips, and at once the greater outside push upon the water fills the mouth and we are ready to swallow. Then, too, in eating, we place the morsel to be chewed between the teeth with the tongue; but how is it removed when the work is finished? If you will watch yourself, you will observe that with both the lips and the passage to the nostrils closed, the lower jaw is dropped, which destroys the balance of pressure on the flexible cheeks, when the air on the outside immediately tucks them in between the teeth, and this crowds the morsel out, all so easily, so unconsciously, and in so brief a time that it is not without difficulty that we can convince ourselves of the real means used to do the work.

It is the unbalancing of this same atmospheric pressure that produces the gentle breeze, and the terrible tornado; that lifts the water in the pump and in the siphon; that produces the draft in the chimney and the in-going and out-going currents of air which constitute soil-breathing or soil-ventilation, so essential to plant life.

I have said that we are living at the bottom of an aërial ocean, and this ocean has a depth of 200 or possibly of 500 miles, but, unlike the one of water, it grows so rapidly less and less dense as its upper surface is approached, that in rising upward from the ground through it one would leave behind him in the first 15 miles all but 4.8 per cent of the entire mass; and yet this thin upper portion offers resistance enough to those stone-like bodies called meteors, traversing space, so that when they plunge beneath its surface, moving at their fearful rate, they quickly become intensely hot or are even converted entirely into gas by the great heat produced during the fall, long before the lower depths of the atmosphere are reached.

Were we to separate the molecules of different kinds which together make up the atmosphere, we should find a large variety of them, but if, after doing so, we were to compare them, volume by volume, we should find, under average conditions, in every 100 cubic feet of air not far from 20.61 cubic feet of oxygen, 77.18 of nitrogen, 1.4 of water vapor, .04 of carbon dioxide, and .78 cubic feet of a recently discovered gas, named argon, reported at the last meeting of the British Association by Lord Reighley and Professor Ramsey, as composing about one per cent of the nitrogen of the air. Professor Morley found at Hudson, Ohio, in his very critical analyses, made in duplicate daily and continued for six months, that the oxygen composed 20.949 per cent of the oxygen and nitrogen taken together, which would make the per cent of nitrogen 79.051.

Besides the substances mentioned above as being found in the air, there should also be named, on account of their important relations to agriculture, ammonia and nitric acid, together with a modified and extremely active form of oxygen, named ozone; and when it is stated that water, carbon dioxide, nitrogen, ammonia, and nitric acid, taken directly or indirectly from the atmosphere, contribute more than 97 per cent of all the materials which are built into the tissues of plants, we can begin to understand how great a part is played by the aërial ocean in the bottom of which we live.

There is another extremely important office which the atmosphere performs, and that is to keep the earth warm. When the sunshine reaches the upper limits of our air, it enters it almost wholly unimpeded, and only as the last few miles of the lower depths are reached is there an appreciable number of its waves turned back into space or absorbed by the air in their passage through it. But when these ether waves break against the land and water surface of our planet, their force is very largely spent in setting the molecules of soil and of water swinging to and fro more rapidly than before, and this means to make them warmer. But to make it clear just how the air is able to save this warmth for us, another fact must be mentioned. It has been said that the ether waves coming to us from the sun set the soil and water molecules swinging when they strike against them. Now the opposite of this is the case at the surface of the sun. There the swinging molecules strike the ether of space, and by that very act lose so much of their power as speeds away to the earth in the waves produced. The rate of vibration of the molecules at the solar surface is, however, very rapid, so that the waves which are sent down through our atmosphere to the earth are very short, so short, indeed, that they seem to make their way among the molecules of our air without disturbing them; but the to-and-fro motion which these waves engender in the earth molecules is very much slower, and hence when these in their turn start waves back through the ether toward the sun, as they do, these waves are much longer and are unable to make their way past the air molecules without setting them swinging; but this means to make the air warmer, it means to hold the sunshine imprisoned for the time being, but in the altered guise of terrestrial heat.

How great this accumulation of heat is will be appreciated when it is stated that Professor Langley, after making a long and very careful experimental study of this property of our atmosphere, at the base and summit of Mt. Whitney, in California, reached the conclusion that, had our earth no atmosphere, its surface temperature, even under the equator at noon, would be 200° C. below freezing, and this means a temperature of $-328°$ F.

With this fact before us, and remembering how slowing down the molecular motion in the tissues of plants, even to the rate of our winter temperatures, stops all plant growth, there is no difficulty in realizing how important this property of the atmosphere is to the life of both plants and animals.

The transformed solar energy does not accumulate at the earth's surface indefinitely, but only until a certain degree of intensity is reached, and then the mean yearly out-go exactly balances the mean yearly in-come. Just how this balance is attained, one will readily understand if we use an illustration from another field. Suppose we have a tall vessel with a hole in one side near the bottom, and that into this vessel a stream of water is flowing over the top. Evidently the in-going stream may be so large that, at first, more water enters over the top than escapes through the opening at the bottom; but there will come a time, if the vessel is deep enough, when the pressure forcing the water out becomes so great as to cause the quantity of water escaping to exactly equal that which enters, and under these conditions a balance is established. So it is with the out-going heat of our earth; its intensity increases under the resistance to its escape by the atmosphere until the jostle of the air molecules among themselves becomes so great that enough waves of the long sort escape into empty space to exactly compensate for those which are being transformed at and near the surface of the earth.

Not all portions or constituents of our atmosphere exert the same screening power over the long dark waves radiated back into space by the earth. It is the lower layers of the air, and especially those portions which are heavily dust and moisture laden which exert this power in a pre-eminent degree. And it is because of this fact that soil temperatures decrease as the altitudes on mountain sides increase until, even under the equator, permanently frozen ground and eternal snow-fields and glaciers may be met only four or five miles above sea level. So, too, when the sky clears after a winter storm and the air has been swept exceptionally clean of its moisture and particles of dust by the forming crystals of snow, each of which has its beginning about a particle of dust, that the radiations escape rapidly through the clear air, giving rise to the cold waves which traverse the country in the rear of winter storms. This is also the reason, in part, why killing frosts in fall and spring usually occur only on clear nights; and why it is that the temperature of arid regions falls so rapidly as soon as the sun has set.

Looking at the atmosphere once more in its relation to life, we find that it performs a very large and very important work as a distributing agent or means of transporting the food and waste of every living being. Taking up the carbon dioxide as it is thrown off in the soil by the germs which consume dead organic matter there, from the lungs and tissues of animals living both upon the land and in the water, from the craters of active volcanoes, and from the fires kindled by man, it is borne by the winds into contact with the green parts of plants, where its carbon is appropriated in the processes of growth. Taking up the oxygen, too, as it is set free from the carbon dioxide in the tissues of plants, the winds bear it back to the soil, to the tissues of animals, and to the fires of the home and the workshop. Taking up the water, too, of which the plant must use from 200 to 300 pounds for every pound of dry matter produced, it is borne fresh and sweet from the salt sea and deposited upon the land, into which it sinks, to be drunk by the roots of plants or brought back by gravity, in the form of springs, to quench the thirst of animals.

WATER AND ITS WORK.

Water is another agent which plays an extremely important part in the processes going on in the soil, and it will be helpful to us if, before entering upon details, we can get a broad view of the work it has done and is now doing.

Studies, both in astronomy and in geology, point to a stage early in the history of the earth when the temperature of the solid land was very far above even a red heat. This being true, there must have been a time when the great ocean sheets which now cover three-fourths of our globe to a mean depth of 16,000 feet did not exist as such, their waters then floating in the form of vapor, shrouding the whole earth in an atmosphere of great density. At this stage water began its great work which to-day is still in progress. Let us see what it has been.

If the small amount of water vapor and dust particles existing to-day in our atmosphere play so important a part in holding back the dark radiations from the earth's surface and thus keeping it warm, as we have seen, how extremely opaque in those early days the atmosphere must have been to these long waves, and how slowly must the surplus heat have passed away by this method! And yet the first great work that water did was to greatly hasten the cooling of the earth down to the temperature at which it might become the abode of life. The manner in which it did its work was this: The water in the upper, clearer portions of the atmosphere lost its heat by radiation and, condensing into drops, fell as rain to a lower and much warmer level, where, at the expense of the heat of the region into which it had fallen, it was quickly evaporated, but only to rise once more into the upper regions to send off into space the large amount of heat which had been imparted to it. The vapor of water, being much lighter than air, rose more quickly than heated air currents could have done, and then, on being condensed into liquid drops, returned again far swifter than it had risen, so that the number of journeys made by these water molecules in their mission of cooling the earth far outnumbered those made by the molecules of oxygen, nitrogen, or carbon dioxide during equal intervals of time. Nor is this all, for each pound of water carried with it on every journey many times the amount of heat which an equivalent amount of air was able to bear.

The next great work that water did was to slow down the rate of rotation of the earth upon its axis until, according to the investigations of G. H. Darwin, our day was changed from one perhaps not longer than three hours to its present twenty-four. It is through the instrumentality of the tides, which, acting as a friction brake upon the earth, have steadily slowed its motion down, but much more rapidly in the early days than at the present time; for then, on account of the moon being nearer to the earth, the tides had perhaps thirty-six times their present magnitude.

As soon as large land areas emerged from the sea, then the third great work of water began; a work which, through all this long time, has consisted in dissolving out, in altering, in breaking into smaller fragments, in grinding, and finally in transporting from higher to lower levels the soil, the minerals, and the rocks of all lands.

This third work of water has been a vast one indeed, but just how great our present knowledge does not make it possible to form even an approximate estimate; that it has been very large, every considerable land area bears ample and indisputable evidence. Take the state of Wisconsin as an illustration: Leaving out of the count the vast depth of deposits which together have been grouped as Laurentian, in this state we have, lying above them, a measured thickness of rock fragments exceeding 30,000 feet, built from materials taken from the soils, the coast lines, the banks of streams, and the subterranean waterways of earlier land areas by this never-ceasing action of water. But great as has been the work of water here in sweeping the dust and litter of land sculpturing into the sea, the strata referred to are the accumulations of less than half the years since the work began.

Finally, with the advent of living forms upon our planet, water took up still another very important work; for to both plants and animals it is indispensable. Making up the larger part of their weight, it is the medium in which the chemical transformations essential to the processes of growth take place, if, indeed, it does not play the part of an active and indispensable agent in bringing these changes about. Then, too, water not only takes up and holds in the liquid form those substances in the soil which may become the food of plants, but it is the medium of transportation by which all materials are moved from root to leaf and from the leaf back to the various places where the processes of growth are taking place. In the physiological processes of the animal body, it has a similar and equally important rôle to play; for here, too, it becomes first a solvent and then a medium of transportation by which all food is taken to, and the wastes removed from, the various organs of the body. Finally, both in plants and animals, water acts as a temperature regulator, tending, by its evaporation from the surface, to prevent the tissues becoming too warm, and among many animals this action is very marked; for they have the power, when the body is getting too warm, of sweating or pouring water upon the surface, where it may evaporate quickly and thus cool the body.

LIVING FORMS AND THEIR WORK.

Not only are the various forms of terrestrial life greatly dependent upon the soil for their well-being, but the soil itself throughout geologic time has been wrought upon in many and very important ways, so that a general statement of the work living forms have accomplished and are now doing will be a helpful preliminary to our study of the soil.

All are familiar with the very rapid washing away of soil by the heavy rains which fall upon steep and naked hillsides, and also with the equally marked protecting influence exerted in the prevention of such washing by the roots and close-lying herbage of plants of almost all kinds. By this action land plants hold in check, in an extremely important manner, the destructive power of water and of wind, giving much deeper and far more fertile soils than would otherwise be possible.

On the other hand, however, these same roots hasten the destruction of rocks by growing into their fissures and wedging them apart, and also by corroding and dissolving away the surfaces both of rocks and of soil grains wherever they may come in contact with them. Nor is this all, for living in the soil, chiefly in the surface 14 inches, are great numbers of microscopic forms, which, feeding upon the dead tissues of plants and animals, evolve large quantities of carbon dioxide, nitric and other acids, which in their turn become corrosive agents, bearing off in the water which runs to the sea vast quantities of soil in solution. Mr. T. M. Read has estimated that the Mississippi alone carries annually to the sea 150,000,000 tons of dissolved rock materials, while other streams bear away proportionately large amounts.

But in this world of never-ending change life has done more than to protect the steep hillsides and to hasten the solution of soil and the crumbling of rocks. In addition it has been a great rock-builder and gatherer of mineral wealth. Taking out of the sea-water the lime which the carbonic acid has dissolved and floated to the ocean, the great army of shell bearers and coral builders have, in all the geologic ages, laid down their mantles and frameworks to become the limestones of the world. In favored situations, too, the decay of the organic tissues of plants and of animals has resulted in the formation of gases, which, rising through the water, have precipitated from it iron and possibly lead, zinc, and copper as it was being borne along in the slow coastal currents, thus bringing into rich deposits in a few places these metals so indispensable to the civilization of to-day. Then there are our great beds of coal and peat, our deposits of asphalt and bitumen, and our reservoirs of mineral oil and natural gas, all of which are believed to have resulted chiefly from the decomposition of the tissues of living forms.

Truly, life, working in the laboratories of the lowermost layers of the atmosphere, in the surface few inches of the dark soil and more widely spread in the transparent ocean waters, using direct and altered sunshine as its moving power, has done and is still doing a very great work. And let the farmer never forget that his life work is thrown among not a few cultivated plants and domestic animals, but rather that every farm is inhabited literally by thousands of kinds and millions of individuals, most of them microscopic, it is true, but powerful in their great numbers. Nor can we for a moment think that only those forms which we consciously aim to raise hold vital relations with us; for year by year, as the horizon of our knowledge of the life histories of the living forms about us is made broader, it is only yet again and again that we learn of new and important relations existing between them and us.

OVER AND OVER AGAIN.

The tide comes and goes and comes again. The morning dawns, the sun sets, the stars come out, and once more the sun is in the east. The cold winds cease to blow, the birds come and then are gone, the snows drift high along the fences, but spring is sure to follow. This method of cycles in the ongoings of nature is so general and so fundamental to the constancy of results, especially in agriculture, that we may well pause for a brief general consideration of it.

The critical and quantitive methods of investigation which so strongly characterize the nineteenth century have led us to know that neither the material of things nor the power which does work can be destroyed. No discovery of modern science is more fundamental and far-reaching than that of the indestructibility of both matter and energy, and equally fundamental is the other fact that, do what we will, we can create neither the one nor the other.

We may, if we choose, conspire to have the energy of burning straw, in the fire box of the engine, converted into the energy of steam and transmit this through the piston and driving belt to the thresher where the work is done ; but while this is going on, through the friction of belts and bearings, and that of the agitation of the straw, grain, and air, there reappears ultimately, in the form of heat, an amount of energy equal to that given to the steam which drove the piston while doing the work, and this too is a measure of the direct and altered sunshine required to perform the labor of building the straw used for fuel. Over and over again is energy used to do the work of the world, but in altered form and in divers places, nowhere destroyed and nowhere created.

The few bushels of ashes left after burning the winter's supply of wood seem to point to a destruction of matter, but their weight, added to that of the products which escape through the chimney, is actually much greater than the original weight of the fuel; for much oxygen from the air has united with it. So it is with our domestic animals; what we realize in their increase of live weight, and in the weight of the dung and urine, falls so far short of that of the food consumed that here is a seeming destruction of matter, but when the materials which are thrown off in the invisible form from the lungs and from the skin are taken into the count, the wastes exceed the food consumed by the amount of oxygen which the animal has taken from the air. Now, in both of these cases, the oxygen which the sunshine released from the carbon in the tissues of the growing plant finally returns to reclaim its carbon again and bear it back as carbon dioxide to the atmosphere from which it was taken.

The water, too, falling as rain upon the soil, and rising in the sap to contribute its constituents to the building of the woody fibre of the fuel, or the starch and sugar of the food, is again returned to the atmosphere to make the rounds once more. It is the same way with the nitrogen and with the ash ingredients, — each and all are used over and over again, but nowhere in the round is there any loss of matter or any gain.

Take our atmosphere as a whole: In the equatorial zone of strongest heat the air is steadily rising to a considerable height, where two currents part and move toward the poles, but only to return as under currents and to make the trip over again. Then, besides this world-wide circulation in the atmosphere, there is a tendency, most of the time, for the air to travel from the land to the sea and from the sea back to the land again, the currents being along the land surface from the sea during the warm portions of the year, but back again toward the sea overhead; then when the seasons are cold and the sea is warmest, the air from the land areas slides along their surfaces, out upon the sea, while, as an upper current, it travels back again to return once more; and by these two great systems of winds the land is watered by the moisture brought from the sea, and the general composition of the whole atmosphere is maintained remarkably constant.

In addition to these larger systems of circulation, the greater absorption and transformation of sunshine at the surface of the earth than takes place higher up in the atmosphere, maintains everywhere and at all times local ascending and descending currents, so that, no matter how rapidly vegetation may consume the carbon dioxide of the air, and put in its place free oxygen, or how rapidly animal life in the air, or microbe life in the soil, may use the free oxygen and put in its place carbon dioxide, these local ascending and descending currents keep the air so thoroughly stirred that the most careful chemical analyses reveal only exceedingly small variations in the relative proportions of the oxygen, nitrogen, and carbon dioxide in the air; and since both plants and animals tend constantly to disturb this relative proportion of gases, it is plain that by this over and over again process, a healthful atmosphere, so essential to the well-being of both plants and animals, is maintained.

Let us now try to gain an idea of the magnitude of the movement in the endless round which water makes as it journeys from the sea to the land, and back from the land to the sea again. Observe the face of your watch just one minute, and then reflect that, on the average, during each such interval of time, the Mississippi River empties into the Gulf of Mexico 40 acres of water more than 21 feet deep; while in South America, its great Amazon unloads a burden nearly five times as large, or more than 103 40-acre-feet per minute. But large as this work really is, it does not measure the volume of water which, on the average, is steadily rising from the earth's surface into the atmosphere under the impulse of altered sunshine; for a large part of the water which falls upon the land is evaporated there, to return as rain in another place instead of being carried away by the rivers. On an area equal to the state of Wisconsin, for example, where the mean annual precipitation measures about 3 feet, the total fall must exceed an average of 40 acres of water 5 feet deep, for each minute of the year. But there are large areas of land where the mean annual precipitation is 60 inches, and others still where 8 feet of water fall: In these cases, for an area equal to the state of Wisconsin, or 56,040 square miles, the aggregate precipitation must exceed 8.5 40-acre-feet of water per minute in the first case, and 13.6 in the second. Now, on only a moderately fertile soil, the writer has grown maize, supplying it with water just as rapidly as it could use it to the best advantage, and found, as an average of two trials, that it did take, during its growing season, or one-third of a year, the equivalent of a rainfall of 34.3 inches, and produced a yield, when calculated for an acre, of more than four times a very large field crop grown under the best natural conditions of rainfall in Wisconsin; so that to grow a field of corn, of such quality, and the size of this state, would require the delivery of water upon it at the rate of 40 acres more than 14 feet deep every minute during the growing season, or a rainfall greater than the largest considered above. Large then as this movement of water is, it is seldom great enough during the growing season to enable a moderately fertile soil to produce its largest crops.

But it is neither to the gaseous nor to the liquid portions of our earth that this process of over and over again is limited; for even the solid land is profoundly involved in it. Careful measurement has shown that there goes to the sea annually, dissolved in the waters of the Mississippi River, 150 million tons of rock; and of these, 70 millions are the chief constituents of limestone — carbonates of lime, and magnesia; so that the selfsame materials which journeyed to the sea, dissolved in the rivers of unnumbered centuries ago and laid down there by the action of marine life, have since become a part of the dry land, certainly once, if not many times, and are now journeying back to be rebuilt into the coral reef on the ocean's bottom yet once more.

Nor is it rock and soil held in solution simply which water in its ceaseless rounds is bearing back to the sea; for with the dissolved materials there is borne along as suspended sediment in the waters of the Mississippi alone, in the space of a single year, 362 million tons, making 513 million tons of rock and soil carried to the sea by one river, besides 750 million cubic feet of matter which are shoved along the bottom to form its delta.

The continual transfer of these large amounts of material from the land to the margin of the sea bottom, perpetually destroys the balance in the figure of the earth, so that the land areas rise to compensate in large measure for the materials borne away, while the marginal sea bottom subsides in like proportion to readjust the balance; but as the sea sediments continue to subside, they become plastic under the great pressure to which they are subjected, and flow toward and under the rising land areas where denudation has been going on, so that in a very powerful but extremely slow manner there is a real movement of the solid land toward the sea above and from the sea back again to the land beneath. And if such profound and long-enduring systems of rotation as these are maintained as essential to the life of the world as a whole, we may practice with great confidence a rational rotation of crops in our systems of agriculture.

> "We cannot measure the need
> Of even the tiniest flower,
> Nor check the flow of the golden sands
> That run through a single hour;
> But the morning dews must fall,
> And the sun and the summer rain
> Must do their part and perform it all
> Over and over again.
>
> "Over and over again
> The brook through the meadow flows,
> And over and over again
> The ponderous mill-wheel goes.
> Once doing will not suffice,
> Though doing be not in vain."

CHAPTER I.

THE NATURE, FUNCTIONS, ORIGIN, AND WASTING OF SOILS.

OF all commonplace things, it would be difficult to find one more uninteresting to most people than soil. Walking over it all our lives, it has come to be, in our unreflective moods, simply dirt, something essentially unclean and to be shunned. So deeply ingrained is this feeling that it comes to many almost as an inheritance, and people of culture, as well as the ignorant, find themselves stoutly inclined to shun everything and everybody directly associated with it.

But the spirit and results of investigation, which have grown so rapidly during our century, have already so widened our horizon of knowledge, and so changed the attitude of mind toward the phenomena of nature about us, that we are coming to study, in the spirit of science, many of those things which lie nearest to us, and with great moral, intellectual, and pecuniary profit; and since soil, air, and water are indispensable to all forms of life, we must know more and more of them as the demands for food and homes increase.

Taking samples of soil from where we will, whether from the fertile prairies of central North America, from the tundras of Siberia, from the barren wastes of the Sahara, or the rich river bottoms of the Amazon, everywhere we shall find them composed of mingled fragments of materials of various kinds. Usually the soil is composed chiefly of small fragments of rock of many varieties, which may be regarded as the basis of them all. Associated with these fine rock remnants there is almost always a varying amount of organic matter derived from the breaking-down of vegetable and animal remains. Then, too, adhering to the surface of these fragments, or scattered among them in the form of crystals, there are various substances which have been deposited from over-saturated solutions of soil moisture.

In clayey soil there is present among its fine silt-like particles a small quantity of silicate of aluminum having water combined with it, and which gives to it its sticky, plastic, or putty-like quality. This adhesive clay, however, forms only a small part of the whole weight of such soils, amounting to not more than 1.5 per cent, according to Schlösing. Then, too, the soils in many parts of the world have scattered through them larger or smaller blocks of stone, varying in size from great masses sometimes weighing tons, down through those which a man can barely move, to pebbles and coarse grains of sand. These fragments form no part of the soil proper, but instead are the materials out of which soils are made. It is true that the roots of plants may place themselves alongside of these coarse pieces of rock, and by their action derive some nourishment from them, but the amount thus obtained is insignificant when compared with that which an equal volume of soil might contribute if placed in their stead; so that such rock fragments, while they will contribute their volume of soil to the agriculture of the future, are a positive hindrance to that of the present.

SURFACE SOIL AND SUBSOIL.

In humid climates, where the rainfall is sufficient to insure remunerative crops, it is common to speak of the surface, 6 to 12 inches of the fine rock fragments, as constituting the soil, while the deeper portions are spoken of as the subsoil, and this distinction grows out of the fact that oftentimes when the deeper soil is brought to the surface, it is found to be unproductive for a time, and, besides, there is generally a sharp line of demarkation in the color of the two portions. In arid regions, however, where crops can only be raised by irrigation, both of these distinctions largely or wholly disappear; so much so that, in leveling fields to fit them for an easier distribution of water over the surface, little or no care is taken to avoid exposing the subsoil or covering even deeply the surface layer, experience having proved that earth from the bottom of cellars, and even that from depths of 30 feet, may be quite as productive, if not more so, than that which has been long exposed to the air.

This difference in the nature of the deep soils of arid and humid regions appears to result from a variation either in the abundance or arrangement of the finest of the soil particles which exist in the deeper layers, the deeper soils of humid regions being usually more close in texture, and less easily penetrated by water.

This difference in the texture of the soils of humid and arid regions is not confined to the subsoil, but involves the surface portions as well, so much so that when most soils of humid climates become dry, the surface is very hard and difficult to move, while those of the arid regions of the world are so incoherent that the slightest puff of wind is sufficient to raise a dust, and a wind storm at any time is quite certain to raise great clouds of sand so characteristic of desert regions.

Just why the soils of dry climates should lack in the amount of adhesive materials is not readily explained by unquestioned facts, but the condition appears to be in some way related to the larger amount of lime which Hilgard has shown these soils to contain. It has been abundantly proved by different experimenters that when the salts of lime are added to muddy water, it has the effect of enabling the silt particles to be gathered together or to become flocculated, settling to the bottom and leaving the water clear, while without the addition of lime it would have remained turbid for an indefinite period.

Hilgard has also shown by experiment that while any clay or tough clay soil, after being worked into a plastic mass and allowed to dry, acquires a texture of almost stony hardness, if to another portion of the same mass only half a per cent of caustic lime be added, a difference in the degree of plasticity is at once observable, and on drying the whole falls into a pile of crumbs at a mere touch; and this change is assumed to result from the expulsion of the water of combination, from the grains of colloid clay, and the gathering of them together into compound particles of larger size, which then lose their cementing power.

We do have, however, many clays very impervious to water, becoming, when worked, quite plastic and adhesive, but which cannot be used for brick or pottery on account of the large amount of lime they contain, which slacks after firing and by its expansion fractures the ware into which it has been shaped. Chemical analysis shows such clays to contain in some cases as high as 5 per cent of lime and yet for some reason the clay has not lost its plastic character; the lime has not produced the flocculation it sometimes does.

Schlösing found, when he was trying to wash a soil with which he was working, until water would pass away from it clear, that, by passing a stream of carbon dioxide through the soil, it had the desired effect, and he attributed the clearing of the filtered water to the formation of more soluble lime carbonate, which, by coagulating the fine clay passing the filter, causes it to be formed into compound clusters too large to escape. In view of this fact it may perhaps be urged that the lime once in the impervious clay could not be acted upon by the carbonic acid sufficiently to dissolve enough to do the work of flocculation, and also that soils more open, as those of arid regions must be when very dry, give easier and more complete access to both the carbonic acid and to what water does fall, so that with the usually relatively larger amount of lime present, owing to the less leaching in dry regions, enough would be dissolved to make a more complete coagulation than generally takes place in the soils of the more humid regions.

RELATION OF THE SOIL TO ORGANIC EVOLUTION.

But soil is very much more than a mass of broken and weathered fragments of inert rock, among which are strewn a small amount of the fast-decaying remnants of plant and animal life. To appreciate the mechanism of that great locomotive which in six days places the fruits of California on the tables of Boston, one must look at it, not cold and still, as so many nicely fitting pieces of

FIG. 1.— Showing surface denuded of its soil by glacial action and not yet re-covered.

polished brass and steel, but alive, the great heart throbbing, the strong arms at the wheels, and the monster starting, hurrying, halting, with its tons and tons of burden, always responding with the utmost promptness and the greatest exactitude to every beck and nod of the intelligence at the throttle. He must realize how, in its furnace open to the free air at both ends, the strength of forty horses is brought out of lifeless coal and placed in a chamber without an entrance doorway and only a chance to escape by doing work against the great piston heads. So if we will understand the soil, as farmers should, we must see it in action, helping on the work of the whole world as well as in producing the basket of apples, the bushel of wheat, or the pound of pork.

How indispensable soil is to the life of both plants and animals as they are now constituted, will be apparent at a moment's reflection when we picture to ourselves the conditions which would exist were all the soil of the entire land area swept into the sea, leaving the surface with the appearance shown in Fig. 1. Under these conditions it is evident that all upright types of plants would be without means for maintaining that position, and there would be no provision, as these plants are now constituted, whereby they could be supplied with water except during times when the rains were actually falling; for the water would hurry swiftly from the surfaces of the naked rock into the main waterways and off to the sea.

Were the land without soil, the vegetation of these areas could only consist of a meagre growth of such forms, among living species, as now subsist upon the naked rocks of mountain sides and similarly exposed situations, where, for any reason, soil is not permitted to accumulate; forms which, like the lichens, algæ and fungi, are not provided with true roots, and which derive almost their whole nourishment, including water, directly from the air or from dead or living organic matter.

Under such conditions as these it is plain that, for lack of food, if for no other reason, there could be no such profusion of terrestrial animals as dwell in a land of plenty among us to-day. It is true that the tundras of arctic climates produce comparatively heavy crops of lichen growths such as the "Iceland moss," which, it is said, often forms the sole food of the poor inhabitants of that lonely land, like the "reindeer moss," which in northern Europe and in Siberia is the chief food of the reindeer and, in times of scarcity, ground and mixed with flour, that of man as well, and like the "Tripe de Roche," eaten by Indians and Canadian hunters in arctic North America. And then there is the "manna lichen" growing on the arid steppes of the countries between Algiers and Tartary, which in times of drought and famine is used as food for large numbers of men and their domestic animals. But the luxuriance of growth in these cases, although small when compared with that of other vegetation, is larger than it could be did it not grow lying upon the soil which holds the water to be given to the air about them by the more gradual process of evaporation as they need it.

For not only does the soil make possible a very much greater profusion of land life than could otherwise exist, but it has also played an extremely important part in that long-continued, never-ending, and sublime process of evolution whereby, as lands have insensibly changed into sea and seas into land, as mountains have risen so slowly and silently out of level plains as to spring their broad arches directly across wide rivers to the height of a mile and yet leave their courses unaltered, as climates have changed from cold to warm or from wet to dry, both plants and animals in this great drama of world action have been enabled to change, not simply their costumes, but if the exigencies of the new scene demanded it, legs for fins or even abandon them altogether and crawl upon their bellies through the grass.

As the soil slowly became thicker and thicker, as its water-holding power increased, as the soluble plant food became more abundant, and as the winds and the rains covered at times with soil portions of the purely superficial and aërial early plants, the days of sunshine between passing showers, and the weeks of drought intervening between periods of rain, became the occasions for utilizing the moisture which the soil had held back from the sea. These conditions, coupled with the universal tendency of life to make the most of its surroundings, appear to have induced the evolution of absorbing elongations, which by slow degrees and centuries of repetition came to be the true roots of plants as we now know them.

When plants came to have specially organized absorbing surfaces placed in a supply of moisture which periodic rains made practically permanent and bounteous and which long contact with the finely divided soil grains kept continuously supplied with plant food not obtainable before, except in the most meagre quantities, the natural consequence to follow was a much more vigorous and larger growth of the parts above ground.

But as bounteous feeding pushed the parts of plants higher above the surface of the ground, they were brought where they were obliged to withstand a much stronger wind pressure than when a precarious supply of moisture kept them close to the surface, and hence, in order to survive and utilize the new opportunities which a fertile soil affords, it became necessary to develop a stronger and more rigid tissue than the lower types of plants possess, and woody fibre in its various forms was the result.

And then, with a deep, rich soil in regions of frequent and bounteous rains, with roots spreading wide and deep to gather and lift the percolating water, and with the woody structure of stem fixed, there began that race for sunshine which has led on and on from low to taller forms, each contestant in the battle striving always to lift and unfold its leaves in the sunshine and free air above all competitors, until there has resulted a vast array of forest trees culminating in the giant Sequoias of our Pacific coast, some of which have attained a measured height exceeding 20 rods or 352 feet.

The soil has made possible succulent, nutritious grass, great forest trees and flowers with beautiful petals, fragrant odors and sweet nectars, while, with the slow evolution of these forms, there has come into being, to use them and to contribute to their welfare in return, the cattle and horses of the plains, the birds and squirrels of the forests, and the bees and butterflies which, guided by the colors and the fragrance they have learned to know, reach the nectar the flowers have provided for them that they shall be sure to come and distribute the pollen and secure to the plants, by cross-fertilization, that renewed vigor so essential to them.

THE SOIL A SCENE OF LIFE AND ENERGY.

In the agricultural sense it should be observed that the most important use of soil is to act as a storehouse of water for the use of plants, and that the productiveness of any soil is determined in a very large degree by the amount of water it can hold, by the manner in which that water is held, and by the facility and completeness with which the plant growing in it is able to withdraw that water for its use as it is needed. But while this statement is true in the fullest sense, it must not for a moment be thought that the composition of the soil is not an important factor in fixing land values for crop production. The great importance of the water-holding power grows out of the fact that without an adequate supply of water, neither the other food constituents which the soil contains, nor that larger part which is derived from the air, can be procured by the plant, nor transformed or assimilated by it.

Then, again, the soil is a wonderful laboratory in which a large variety of the lower microscopic forms of life are at work during those portions of the year when its temperature is above freezing, breaking down dead organic matter and converting it into those forms in which it again becomes available for plant food; and the farmer should never forget that the crop of these invisible organisms which are produced each year in his soil, determines in no small degree the magnitude of the harvest he removes from the ground and the fitness of that ground for a succeeding crop.

Finally, the soil is a means for transforming sunshine and putting it into a form available for carrying on the kinds of work which are there accomplished, and the manner in which the soil is tilled and the way it is fertilized have much to do with the quantity of altered sunshine which becomes available in carrying on this work.

INFLUENCE OF ROCK STRUCTURE IN SOIL FORMATION.

There are many agencies at work in the production of soil, and the process is one which is being carried forward continuously night and day and almost incessantly the

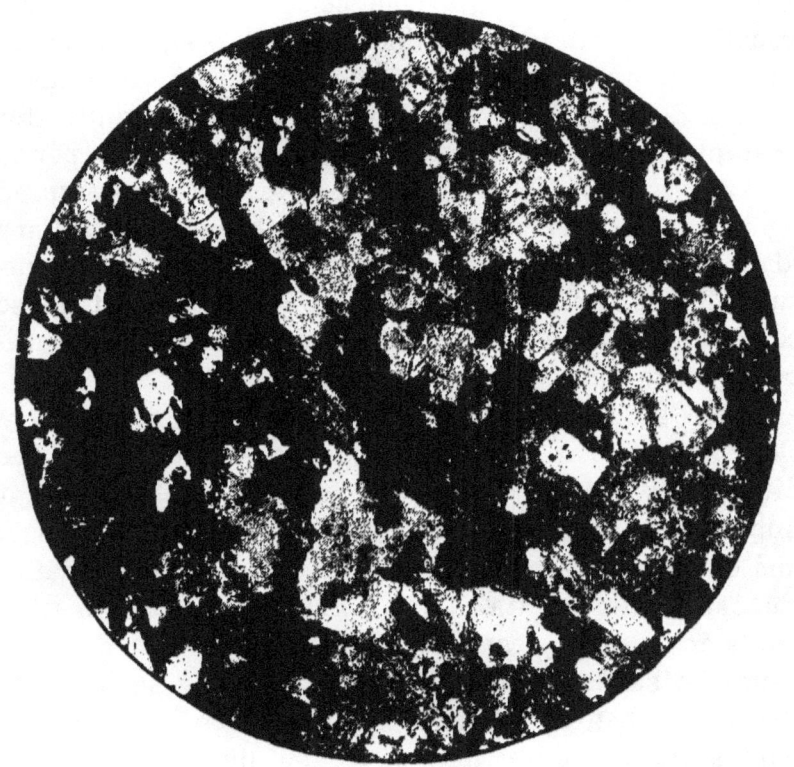

Fig. 2. — A thin section of mica schist, showing the crystalline structure which lends itself to the conversion of rock into soil.

year through. It is taking place in the tilled field and in the meadow; in the depths of the forest and on the prairie; in the driest desert regions of the world and under the tropics where rains are of daily occurrence; in the polar regions and under the equator and on the topmost summits of mountain masses as well as along the margins of the lake and the sea, or where streams at times rise and overflow their banks.

Nearly all rocks are made up of fragments or crystals of various sizes and kinds, and these are bound together more or less firmly by some cementing material, but usually there are places which have not been completely filled with the cement, and these give to most rocks a certain degree of permeability to water. Granite, for example, has been found to absorb nearly .4 of a pound of water to each 100 pounds of rock, and the fine-textured agate is open enough to admit of coloring by capillary absorption. Fig. 2 conveys a general idea of that structure of rocks here referred to, and which lends itself so readily to their conversion into fine fragments and ultimately into soil.

When such rocks are brought to the surface, where they are exposed to wide ranges of temperature, the different rates of expansion possessed by the different minerals entering into their structure tend to loosen them and open the natural cavities wider. Into these cavities the rain water is drawn by capillarity, and on freezing tends to open the cavities still more, and to flake off from the surface minute fragments, adding so much to the soil. Then as the rocks by this action become more porous, the rain water, holding carbon dioxide in solution, enters more freely and dissolves the more soluble minerals, and, as the surface dries, these dissolved materials are brought out and washed away by the rain or blown off by the winds, thus leaving the rock more and more porous until finally it falls into fragments, illustrated so well and so frequently by what are popularly called "rotten stones."

In moist and warm climates the solvent power of water here referred to, is exerted with the greatest vigor, but in damp, cold countries the effects of freezing are most apparent, while in desert regions neither the one nor the other become soil-producing agents of any note.

Fig. 3. — Showing the conversion of rock into soil on a limestone hill.

Whoever will visit an abandoned stone quarry where the rocks have lain undisturbed for ten or twenty years, will readily observe in the stained and altered surfaces, in the softer, more easily scratched outer layer, and in the slight accumulations of soil, which the rains and the

Fig. 4. — Showing the transition from rock to soil on a limestone plane.

winds have swept into the small inequalities of the rock, the initial process of soil formation as it is still taking place and has been throughout geologic ages.

Passing from the abandoned quarry to some fresh cut along the railway or roadside, where a hill has been graded down, there may be seen at the top the finished soil, and between it and the unaltered glacial gravel or original rock below, as the case may be, every degree of progress from the one into the other, represented in Figs. 3 and 4, where the first shows the stages of transition as

Fig. 5. — Showing advanced state of erosion. Giant's Castle near Camp Douglas, Wisconsin.

they have taken place over the summit of a limestone hill, while the second shows the same facts for a more level limestone surface.

Then, again, the rocks of almost any quarry, on examination, will be found to be divided into blocks of varying size and form by fissures or breaks, which owe their origin to a general shrinkage of the surface layers and to small but ever-present bendings or wave-like motions of the ground. These features are well brought out in Figs. 5 and 6, and they introduce us to another process in the formation of soil.

Fig. 6. — Showing the last stage of the conversion of cliffs into soil.

Such fissures as these, when not too deeply covered with soil, are often penetrated by water and by the roots of trees. Then as the water freezes and as the roots grow, both expand with an almost irresistible power and open the old crevasses wider or make new breaks where none existed before, thus dividing the larger blocks into smaller ones, and often throwing the fragments to the foot of the cliff, where they soon become overspread with a mantle of vegetation under which they rapidly fall into soil. This process as it is carried forward in nature may be better appreciated by carefully studying Fig. 7.

Those who live near the foot of great rocky cliffs which are subject to this sort of action of ice, like the quartzite cliffs at Devil's Lake, Wisconsin, are frequently startled during cold nights in winter by loud reports followed by the sound of rolling stone as great blocks of rock, sometimes many tons in weight, snap under this action of frost and go bounding down the steep face of the pile of angular fragments which have accumulated at the foot under the same action many times repeated. Then, again, where a great tree has grown up with its roots reaching deep into some wide fissure filled with soil, on the summit of an overhanging or vertical cliff, it not unfrequently happens that a strong wind from the right direction, pressing against the wide-spreading boughs and using the tall trunk as a lever, pries off sections of the cliff, sometimes 20 feet long, as many deep, and 6 to 10 feet thick. Such a block the writer has known to fall during a wind storm.

RUNNING WATER AS A SOIL BUILDER.

Running water is another agent which, by processes peculiar to itself, has done very much in the production of soil and in giving to it certain characteristics. Standing on the bank of any small stream and watching the water as it slides over the bottom, it will be seen that there are incessantly being moved along the bed, some-

FIG. 7.—Showing the crumbled rock fragments thrown down from cliffs and passing into soil.

times rolling, sometimes sliding, grains of sand of varying sizes. One set moves on for a short distance and then stops, other grains follow after but halt at the same place until a small but appreciable ridge is formed. Similar tiny ridges are forming on its right and on its left, some above it and others below. But these are only brief resting places; for a change in the velocity of the current as the waters shift from side to side causes each ridge to move another step farther down the stream. But as the grains roll, tumble, and slide by turns in their downward course, each has its corners worn away, each is growing insensibly but surely smaller, and each is contributing something to that impalpable powder, which, rising into the body of the stream, remains suspended for almost indefinite periods except in the stillest water, where it is laid down in lakes or carried to the sea to slowly subside and become beds of clay, and when those geographical changes come which drain a lake or elevate the margin of the sea bottom into dry land, it then becomes fields of clayey soil.

But what becomes of the grains of sand which are only moving along step by step, and where was their resting place before they joined this caravan traveling toward the sea? A little observation will soon show where their journey begins. Following along the bank until a bend in the stream is reached, it will be at once observed that the concave side of the channel is deep, while the convex side is shallow. On the concave side the current is swift and plainly cutting away the bank, which now rises abruptly above the water, or perhaps has come to be so far overhanging that a portion has already broken loose and slid into the stream, where the rapid current is sorting out the grains it is able to move along and bearing them beyond the turn. But on the convex side the ground slopes gently almost to the water's edge; the current is feeble, and hence the grains of sand advancing from above are here being laid down to fill the channel on this side as rapidly as the stream cuts away the bank on the other. The bed of a stream, then, is constantly shifting; soil is being taken from one side and borne along for a distance and then laid down upon the other, where it constitutes a new soil, and will remain as such until the stream sweeps back once more across its valley. Looking at Fig. 8, which is a map of a portion of the Mississippi below Vicksburg, it is plainly shown how the action at the bends here referred to is being carried on, the dotted areas in the channel showing where the deposits are taking place, and the clear portions where the banks are being crowded back. There is one thing about this map, however, which it is not so easy to comprehend, and that is the amazing extent to which this great river is to-day sauntering about upon its alluvial plain, some 40 miles in width, as shown by Lakes Bruin, Palmyra, and St. Joseph, which are only great ox-bow portions of its channel which it has recently abandoned. Colonel Abbot says of these shiftings of the river: "Chief among such changes is the formation of cut-offs. Two eroding bends approach each other until the water forces a passage across the narrow neck. As the channel distance between these bends may be many miles, a cascade, perhaps 5 or 6 feet in height, is formed, and the torrent rushes through it with a roar audible for miles. The banks dissolve like sugar. In a single day the course of the river is changed, and steamboats pass where a few hours before the plow had been at work." The

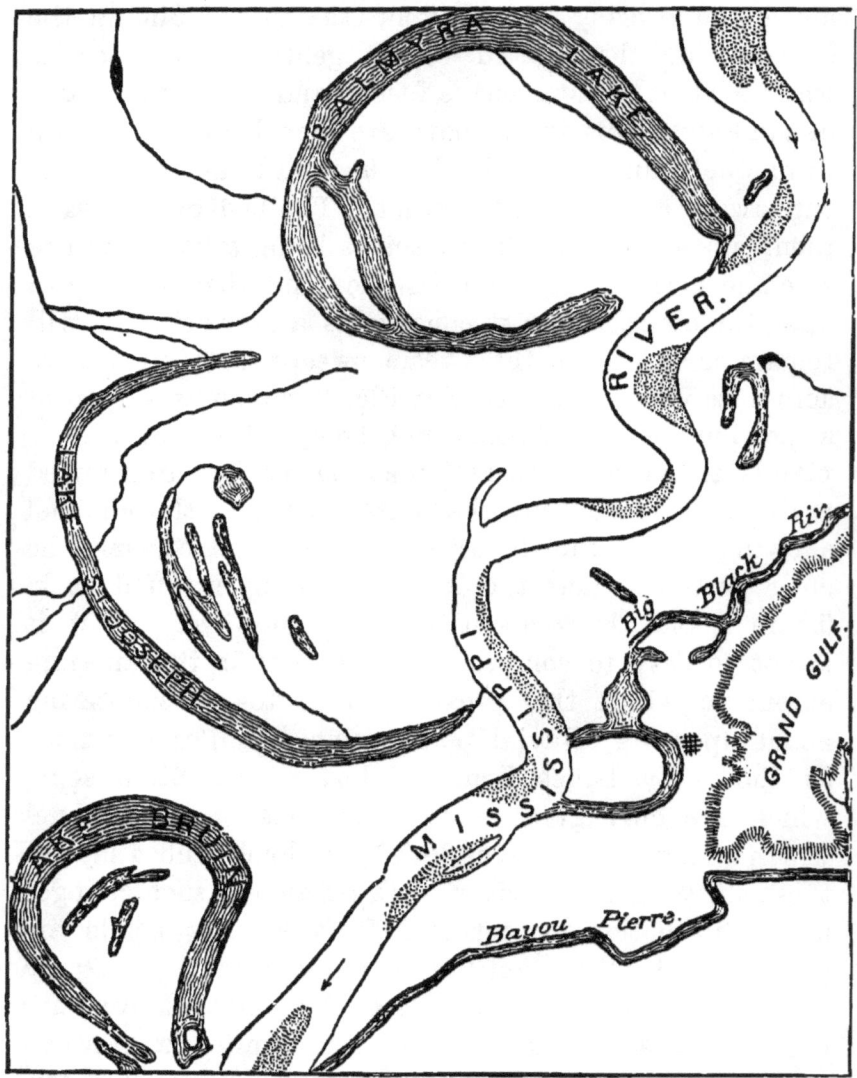

Fig. 8.—Showing the shiftings of a river channel as it forms alluvial soils.

work here is of course extraordinary, as it must be when the run-off from more than 1,200,000 square miles is brought together into one channel on a plain which falls only one inch in 40 rods, and over which it winds some 1100 miles in making a distance south of less than half that number.

While such extreme windings as these are confined to large rivers where they traverse very flat stretches of country, this shifting of the streams, this picking up of soil from one place and laying it down in another, is nevertheless very general and very extensive; and when we speak of the Mississippi as carrying to the Gulf, suspended in its waters or shoved along its bottom, every year soil enough for 72 sections of land 4 feet deep, this work is small when compared with that which measures the shifting of sand from side to side by the same stream even after it passes the city of Memphis; and vast as this work can but be, it must constitute a standard all too small by which to measure that which in the aggregate is done throughout the broad valley of the Ohio, the Upper Mississippi, and the Missouri, with their tributaries. Glance for a moment at Fig. 9 that it may be realized, not only how many times the Madison fork of the Missouri River must have crossed and re-crossed its broad valley, but how many times over and over again it must have handled that soil before it succeeded in carrying out of its field of labor the amount which the unimpeachable testimony of those terraces show it has transported to lower levels. Think also how, again and again, a new vegetation must have taken possession of the reworked land, as the soil was being transferred now to the right bank and now to the left. Years and years, and even centuries, must have passed before a given field was entirely replowed, resoiled, and reseeded; but everywhere in the past and

FIG. 9.—Showing the windings of a stream and the formation of river terraces. Madison River Valley, Montana.

everywhere in the present this long-time rotation has been and is now being carried on by the ceaseless action of rivers.

WORK OF RAIN.

Besides the action upon the soil, of running water, which we have considered, after it has reached the perennial waterways, there is a large mechanical work performed by the rain while on its way over the surface toward the streams or before entering the ground to emerge again in the form of springs. At the time of heavy rains the surface of the ground, even on rather steep slopes, becomes covered with a sheet of water, and the raindrops, striking into this and upon the soil, work up the looser grains so that, as the water moves down the inclines, it bears along with it over shorter or longer distances a considerable body of soil. In cultivated fields, especially where the soil has a fine texture and the property of losing its coherence when flooded with water, the amount of soil moved at such times on the more rolling lands is very large indeed.

Whoever has the opportunity to traverse the uplands in the state of Mississippi to-day will be confronted with a gigantic but sorrowful example of what this action of rain may accomplish in the brief period of thirty years on such soils as are here referred to. I quote the language of W. J. McGee, who has made a careful study of this region: —

" With the moral revolution of the early sixties came an industrial revolution; the planter was impoverished, his sons were slain, his slaves were liberated, and he was fain either to vacate the plantation or greatly to restrict his operations. So the cultivated acres were abandoned by thousands. Then the hills, no longer protected by the forest foliage, no longer bound by the forest roots, no longer guarded by the bark and brush dam of the careful overseer, were attacked by raindrops and rain-born rivulets and gullied and channeled in all directions; each streamlet reached a hundred arms into the hills, each arm grasped with a hundred fingers a hundred shreds of soil, and as each shred was torn away, the slope was steepened and the theft of the next storm made easier.

"So, storm by storm and year by year, the old fields were invaded by gullies, gorges, ravines, and gulches, ever increasing in width and depth until whole hillsides were carved away, until the soil of a thousand years' growth melted into the streams, until the fair acres of ante-bellum days were converted by hundreds into bad lands, desolate and dreary, as those of the Dakotas. Over much of the upland the traveler is never out of sight of glaring sand wastes where once were fruitful fields; his way lies sometimes in, sometimes between, gullies and gorges, the 'gulfs' of the blacks whose superstitions they arouse, sometimes shadowed by foliage, but oftener exposed to the glare of the sun reflected from barren sands. Here the road winds through a gorge so steep that the sunshine scarcely enters; there it traverses a narrow crest of earth between the chasms, scores of feet deep, in which he might be plunged by a single misstep. When the shower comes, he may see the roadway rendered impassable, even obliterated, within a few minutes; always sees the falling waters accumulate as, viscid brown or red mud torrents, while the myriad miniature pinnacles and defiles before him are transformed by the beating raindrops and rushing rills so completely, that when the sun shines again he may not recognize the nearer landscape.

"This destruction is not confined to a single field or a single region, but extends over much of the upland. While the actual acreage of soil thus destroyed has not been measured, the traveler through the region on horseback daily sees thousands or tens of thousands of formerly fertile acres now barren sands; and it is probably within the truth to estimate that 10 per cent of upland Mississippi has been so far converted into bad lands as to be practically ruined for agriculture under existing commercial conditions, and that the annual loss in real estate exceeds the revenues from all sources; and all this havoc has been wrought within a quarter century. The processes, too, are cumulative; each year's rate of destruction is higher than the last.

"The transformation of the fertile hills into sand wastes is not the sole injury. The sandy soil is carried into the valleys to bury the fields, invade the roadways, and convert the formerly rich bottom lands into treacherous quicksands when wet, and blistering deserts when dry. Hundreds of thousands of acres have thus been destroyed since the gullying of hills began a quarter of a century ago. Moreover, in much of the uplands the loss is not alone that of the soil, *i.e.* the humus representing the constructive product of water work and plant work for thousands of years; but the mantle of brown loam, most excellent of soil stuffs, is cut through and carried away by corrasion and sapping, leaving in its stead the inferior soil stuff of the Lafayette formation. In such cases the destruction is irremediable by human craft — the fine loam, once removed, can never be restored. The area from which this loam is already gone is appalling, and the rate of loss is increasing in geometric proportion."

It is not necessary, however, to go to the bad lands of Mississippi or the Dakotas for examples of this work; for every careful farmer has witnessed it a hundred times on every hillside of his farm, and has studiously tried to prevent it. But this action of rain is much more general and in the aggregate much greater than has yet been indicated; for it takes place, only in a less intense and less obtrusive way, over the surface of all except swamp and the most heavily wooded lands, always moving the surface soil from the higher toward the lower grounds and ultimately to within the reach of streams.

When the soils of hillsides expand with increasing moisture or with frost, there is a small but sure movement downward, for while the push is equal in all directions, the downward thrust has the force of gravity on its side; and the same movement results when the soil comes to shrink after drying, for it is easier to draw the upper particles down than to pull the lower ones up. Blocks of stone, too, lying upon the slope, expanding in the hot sun and contracting during the night, tend to creep insensibly into the valley. When the fox, and the many burrowing animals of whatever sort, bring dirt to the surface, or when great forest trees are uprooted by the storm, the soil moved is without exception left one step nearer to the sea. It is evident that by whatever one of these methods of creeping the soil is moved, the rate of travel will, other conditions being the same, always be most rapid on the steepest slopes, so that generally this action must result in the soils of the summits or high lands crawling down and upon those of the lowlands, producing an overplacement which gives to the soils of the invaded areas characteristics derived from the rocks whose destruction contributed the material for the overplacing soil. For the same reason, too, the resultant soils will usually be found more fertile and more enduring, as every farmer knows, than that left behind on the more sloping ground. The general facts of soil creeping and of overplacement are indicated diagrammatically in Fig. 10.

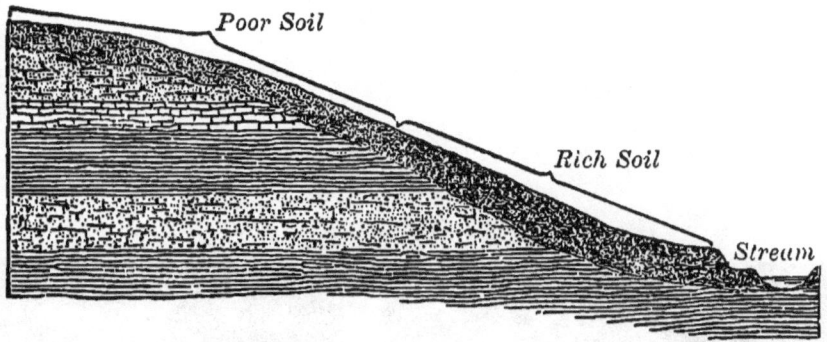

FIG. 10. — Showing movement of soils from higher to lower levels.

GLACIAL SOILS.

In those portions of the world where the temperature is so low that most of the moisture is frozen when it falls and does not all melt during the summer, the snow accumulates often to great depths and by its weight is compressed into ice. When snow in this condition has attained a considerable thickness, it begins to move along sloping surfaces much as liquid water does, converging into larger and larger streams, moving faster where the slope steepens and slackening its speed again when the descent becomes less. In Fig. 11 is shown one of these ice streams descending from the Alaskan mountains toward the Pacific Ocean, where it had its birth.

While the movements of glaciers are much less rapid than those of rivers, their far greater depth and consequent heavy pressure, together with the more rigid nature of the ice, give to these streams a grinding,

Fig. 11.—Showing an Alaskan glacier.

scoring, and transporting power which is great almost beyond measure; and hence it is that in mountain districts, which rise above the line of perpetual snow, and in the frigid zone, glaciers become soil-producing agents of great vigor.

In very recent geologic time, during the glacial epoch, a vast ice sheet gathered in the higher latitudes and spread from the direction of northern Labrador until it overran two-thirds of the North American continent, advancing so far southward as to place its front in the shape of a rude crescent, stretching from Cape Cod and Long Island through northern Pennsylvania into the Ohio valley and from thence, following the course of this stream and that of the Missouri River, to the Rocky Mountains. At the same time there appears also to have been on the Pacific slope a lesser and apparently more local sheet, which pushed itself southward through the mountain valleys until it passed the foot of Puget Sound.

What conditions conspired to induce this long geologic winter has not yet been learned, but during its prevalence the snows piled upon the land until a mantle hundreds and perhaps thousands of feet in thickness overspread the whole area outlined above, while the general level of the ocean fell as its waters were drawn upon to feed the ever-deepening snow fields as they spread over the northern continents of the Eastern and Western Hemispheres alike. As the specific gravity of ice varies between .917 and .922, the mean weight of a cubic foot will exceed 57 pounds, and an ice sheet 10 feet in depth will press upon its bed with a weight exceeding 570 pounds to the square foot, while the burden imposed by 500 and 1000 feet of ice must exceed 28 and 57 thousand pounds to the square foot, or 198 and 396 pounds to the square inch respectively. What a mill for grinding rock into soil we have here! For its nether stone one-half or two-thirds of the North American continent, and for its upper one a block of ice of corresponding size, five hundred to a thousand

FIG. 12.—Showing portion of a terminal moraine near Eagle, Wisconsin.

and more feet thick, having its grinding face thickly set with sand and gravel, and those same hard bowlders, large and small, which we now find strewn so thickly over the surface and through the soil of this whole glaciated area, while the two faces were set against each other with a pressure exceeding 200 to 400 pounds to the square inch and quite probably double these amounts! With such a mill as this, set up under no other roof than the dome of a cold arctic sky, and run incessantly day and night, year in and year out, for centuries, numbered certainly by hundreds if not by thousands, a great work must have been accomplished.

During all this time the great ice sheet was creeping slowly toward the south and southwest, its whole front, from the Atlantic to the Rocky Mountains, now advancing a little and now retreating as variations in the rate of travel or the rate of melting occurred. Beneath the bottom of this slowly moving sheet of pressure-plastic ice, which, with more or less difficulty, kept itself comformable with the face of the land over which it was riding, the sharper outstanding points were cut away and the narrower and deeper river cañons filled in. Desolate and rugged rocky wastes were ground down and overspread with rich soil, and regions with sandstone for the surface rock, which by decay in place could give only lands of the lighter type, became mantled with thick layers of mixed gravel, sand, and clay, forming by slow alteration rich and enduring soils.

Great streams of water emerging from the melting ice sorted and resorted the glacial grist, leaving in some places extensive beds of coarse, clean gravel with surfaces sloping gently in the direction of discharge, over which has since developed one type of extremely fertile prairie

FIG. 13.—Showing a section of a terminal moraine near Whitewater, Wisconsin.

soil; leaving in other places beds of sharp plastering sand, often interstratified in the most curious and abrupt manner with coarser or finer materials, while the finest silt was farther removed to subside in innumerable lakes, formed by glacial dams, or to be borne away to the sea to contribute materials for the extensive deposits of our coastal plains.

When the ice front tarried long at a given place, the

Fig. 14.—Showing glacial scratches on north shore of Kelley's Island, Lake Erie.

broken and worn fragments of rock continually brought along by the ice were unloaded there, as it melted, until great irregular ridges were formed, sometimes several miles wide and from 20 to 400 feet in height, and these have been named terminal moraines. Mingled with these huge piles of rock, sand, and gravel there were left, at the time of the final retreat, great blocks of ice, and when these finally melted and the water drained away, they left the surface of the moraine thickly set with deep and abrupt hollows, to which the name of kettles came to be applied; and Fig. 12 shows, as well as a picture can, a surface thus formed, while Fig. 13 represents a section of one of these morainic hills which was cut through in grading for a railroad.

In very many parts of the country overrun at this time, the harder and more enduring kinds of rock often still show with remarkable distinctness the actual course the ice stream took by the scratches, grooves, and furrows which were left in the polished surface of the rock made by stone, gravel, and grit imbedded in the bottom of the glacier. An illustration of such a rock surface is represented in Fig. 14. And, since these rock surfaces have sufficiently withstood all other methods of rock decay to enable them to retain these records for so long a time as has intervened between the close of the glacial period and the present time, it is evident that glacial action must be ranked as the greatest soil-producing agent of recent geologic time.

EARTHWORMS AS SOIL WORKERS.

Then there are many animals which have contributed largely to rock grinding and soil formation. Many will recall the great number of piles of earth which angleworms bring to the surface after heavy rains, and especially during the spring and early summer when the ground is wet. Their method of action is this: in moving through the soil, they eat a narrow hole, swallowing the earth, when the point of the head is held fast in this excavation while an enlarged portion of the esophagus or swallow is drawn forward forcing the cheeks outward in all directions, thus crowding the soil aside and making the opening wider, when more dirt is eaten and the process repeated. After the swallowed earth has been worked over in the muscular gizzard and the organic matter associated with it digested, it is passed from the body, but while in the stomach the grains of sand suffer a considerable amount of wear, much as is the case with pebbles and bits of glass eaten by poultry and all seed-eating birds for the purpose of grinding their food.

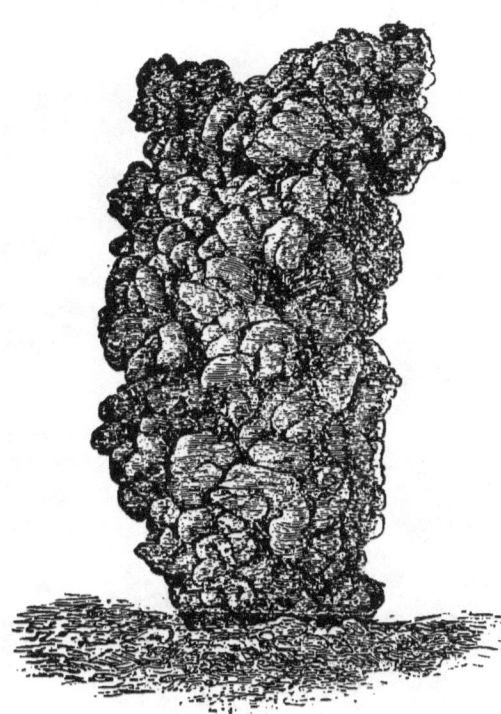

FIG. 15.—Showing a tower-like casting ejected by a species of earthworm from the Botanic Garden, Calcutta. Natural size. Engraved from a photograph after Darwin.

Charles Darwin, who made a very long and careful study of this action of earthworms, came to the conclusion that in many parts of England these animals pass more than 10 tons of dry earth per acre through their bodies annually, and that the grains of sand and bits of flint in these earths are partially worn to fine silt by the action of the gizzards of these animals; but this extensive action is not peculiar to England, for the species of earthworms have a wide geographic distribution, and Fig. 15 represents a soil chimney made by a species in India,

Fig. 16. — Showing the work of the common earthworm during a single night after a heavy rain.

while Fig. 16 shows the work of a single night on a Wisconsin farm after a heavy rain.

HUMUS SOILS.

There is a class of soils in which humus is the dominating characteristic, and these have their origin in swamps of various types. In many parts of the world, but especially in high latitudes or at considerable elevations, where the surface is too flat for rapid and complete drainage and where the winter snows remain so long that the duration of summer is insufficient to dry the soil enough for it to become readily penetrated by the air, there comes to be formed a deposit of humus or peat, mingled with varying proportions of mineral grains. When soil is allowed to become dry and at the same time porous, so that air has an unlimited access to it, all organic matter which it may contain is rapidly and completely returned to the atmosphere, as is illustrated in an emphatic manner in the use of the dry-earth closet; but when organic matter falls upon an over-saturated soil, or beneath the surface of water, there is so slow an access of oxygen to the decaying matter that only a partial decomposition takes place, which results in the formation of the black or brown humus so characteristic of swamp soils. It frequently happens during the development of soils of this type that heavy forest growths are slowly drowned out by it as it accumulates in thickness and more and more completely excludes the air from the roots of the trees.

Under other conditions, where a great river like the Mississippi, the Nile, or the Ganges approaches its outlet or winds over a flat plain, it so builds up the bottom of its channel, by laying down the sediments which it carried with ease over the steeper portions of its course, as to place it higher than the adjacent flats; and then, in times of frequent overflow, the adjacent country is periodically flooded, and the water, not finding ready return to the river, furnishes the conditions for the growth of aquatic plants and a swamp soil begins to form. Then again when rivers change their course, as in the case of the formation of ox-bows already referred to, there result shallow lakes with muddy bottoms in which aquatic plants are quick to spring up and begin the formation of a rich humus which in later years may lead to no little wonderment as to how such a soil in such a shape could have had its origin. The swamp soils of river origin, when they are drained and dry, on account of the sand and silt mingled with the humus, fall among the richest, most enduring, and quickest to respond to agricultural treatment.

If one will study a detailed map of almost any portion of North America lying north of the great glacial front, to which we have referred, he will be surprised at the very large number of small lakes there shown; and then if he could see a good soil map of the same section, he would be still more surprised, not only at the even greater number of small and large irregular isolated areas of swamp soil, but also at the extent of the fringes of these soils, which are shown bordering most of the lakes. The isolated areas of swamp soil were once lakes themselves, which, after partial drainage and the formation of marginal muddy flats by the wearing down of their outlets, began the development of humus first over the flats through the growth of aquatic plants rooted on the bottom, and later, as the deeper water was approached, by a rapid spread of these and the sphagnum moss extending around the lake as a floating fringe. As this sphagnum continues to grow above and die below, with only partial decay, it settles deeper into the water and finally rests upon the bottom. Or if the lake is deep, the growth may advance steadily toward the centre until finally the whole is overgrown. Upon this raft of aquatic vegetation the sedges begin to grow, small willow and heath-like plants follow, and finally, when a sufficient thickness has been formed, the tamarack disputes the ground and covers the whole with a forest of straight green spires. As these trees grow to maturity

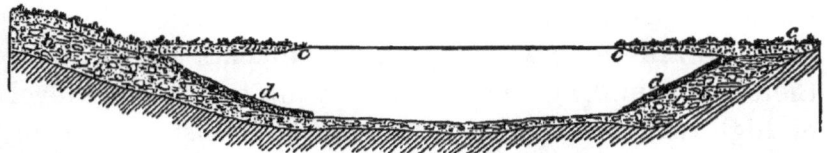

Fig. 17. — Showing the passage of a lake into a peat bed and swamp or humus soil.

and the tall shafts fall to be buried in shrouds of antiseptic sphagnum, the whole lake cover grows thicker, heavier, and finally comes to rest upon the bottom. Fig. 17 shows a section of a lake passing through the stages here described.

It not infrequently happens in these lakes, during the overgrowing period, that large areas of the floating peat break away and drift from side to side, impelled by the wind acting upon the vegetation as a sail. Such a detachment, consisting of several acres, occurred in 1875 in Lake Butte des Morts, through which the Fox River passes before entering Lake Winnebago in Wisconsin. This field of swamp grass drifted against a long railroad bridge, threatening its destruction, and had to be cut with ice saws into blocks small enough to clear the waterways.

On the shores of the larger lakes and along the margins of the sea, the waves often throw up bars of sand, cutting off lagoons whose waters, in the case of the sea, by the shutting out of the tides, come to be converted first into brackish and then fresh-water lakes and finally into areas of swamp soil. And then when the sea has a flat or gently sloping tidal plain over which the water rises more than one or two feet, deposits of peat begin to form in the sheltered bays through the growth of certain grasses which have acquired the power of thriving in salt water. Starting at the shore line, they advance by degrees toward the sea as the tidal sediments gather about their feet, building the bed of peat up to near the level of high tide, and in this way forming extensive deposits. This work is very materially assisted, as Shaler has shown, by the occasional breaking-up of these peat deposits through the action of waves and the redepositing of them in deeper water, where they constitute a soil upon which a peculiar flowering plant, the eel grass, has acquired the habit of living entirely submerged beneath the water. This plant, playing the same part in the deeper water below low-tide mark which the grasses do above it, gathers silt and builds up a platform upon which the grasses may advance in their work of winning land from the sea; and the amount done since the glacial period, along the New England coast alone, is placed by Shaler at more than 350,000 acres, while the total area of marine marshes, which owe their formation to grass-like plants, the same writer places at nearly 10,000 square miles in the United States. But the most interesting and important fact about these lands is, that, like those of Holland, when won from the sea and made fit for tillage, they become extremely rich and enduring soils. And then to think that all this potential wealth of food, happy, industrious homes, and moral lives should lie unused in the very shadow of a great city where so many thousands are worse than starving, where so much wealth is seeking profitable investment, and where so many charitable hearts are yearning to see some pathway leading away from those dens of misery! Why may not intelligence, wealth seeking investment, charity and poverty join hands in converting those great stretches of waste land close by their doors into the richest acres, where the poor who will may not only find employment, but earn a garden and a home, with all these can mean?

WIND-FORMED SOILS.

There is still another method and another agency by which soil building and soil wasting are carried on. It is probable that nowhere can soils be found which do not contain larger or smaller amounts of wind-borne particles, — particles which have traveled unknown distances, and these often very great. There is never a raindrop falls, never a hailstone reaches the ground, and never a snowflake, however white, but which brings to the soil it moistens one or more particles of dust. The streaks of dirt left upon the window pane with the melting of snow which has drifted against the glass are a sufficient proof to the housewife of the truth of the statement just made, while the color of rain water and the sediment which both it and the water of melted snow leave when evaporated to dryness, bear witness to the same facts.

But it is in arid regions more particularly and along sandy coasts, that this action of the wind as a soil builder and soil destroyer is most marked. On the leeward margins of arid regions sand-drowned forests, as well as cities and ancient monuments, stand as silent witnesses of this abraiding and transporting power of the wind. The most extensive formations ascribed to the action of winds are some types of a deposit which have been named loess. Richthofen describes a formation of this character in China as a wholly unstratified calcareous clay of a very fine texture, which contains many land shells, bones of land animals, and land vegetation. The formation as described by him has a wide distribution, and attains a thickness in places of 1500 and possibly 2000 feet. It should be stated here, however, that formations very similar to this have a much wider geographic distribution, being found extensively both in Europe and North America, but these are so related to ancient glacial margins and channels of overflow as to suggest, for them at least, an apparently different origin.

When we call to mind the experience we all have had regarding the accumulations of wind-driven dust which collect in abandoned houses and in rooms even when unoccupied for only brief periods, when we recall the clouds of dust raised from the road and from the surfaces of light, dry soils, or watch the snow as it is driven from field to field, we need not be surprised if the wind has done a great work where it has had an unimpeded sweep and the years of geologic time at its disposal.

CHAPTER II.

TEXTURE, COMPOSITION, AND KINDS OF SOILS.

We have seen that almost everywhere the land surface is overspread with a mantle of rock fragments, sometimes large but mostly very small, the upper portion of which has been designated the soil. We have pointed out that this soil is a water reservoir in which the rains are caught and held, that it is a great laboratory in which certain essential plant foods are being made, and that deeply into it the roots of plants grow for support, moisture, and nourishment.

SOIL TEXTURE AND ITS INFLUENCE.

In looking at the texture of the soil, so far as the size of its grains is concerned, we may consider the valuable results obtained by Hilgard in his mechanical analysis of some Mississippi soils. By his method of treatment he has so separated the grains of soil of different kinds as to bring those having the same diameters together, thus enabling him to determine the percentage amounts of each size entering into the particular soil. The diameters of the soil grains and the percentage amounts of each size, as they occur in a sandy soil and subsoil from the long-leaf pine plateau in Smith County, are given below: —

Soil Texture and its Influence. 71

DIAMETER.				SOIL. PER CENT.	SUBSOIL. PER CENT.
3937 to 4724 hundred thousandths of an inch				.4	.4
2362 to 3937	"	"	"	3.0	.8
1575	"	"	"	6.9	6.3
1181	"	"	"	8.1	3.4
610	"	"	"	3.0	3.9
472	"	"	"	1.6	1.5
283	"	"	"	1.2	.6
185	"	"	"	3.6	2.6
142	"	"	"	6.8	5.4
98	"	"	"	14.6	7.9
59	"	"	"	14.8	17.0
31	"	"	"	30.7	38.3
4	"	"	"	4.6	10.9

In two other soils, one a "Hog-wallow" subsoil from Jasper County, very clayey and difficult to work, and another a loess soil from Claiborne County, we have diameters and relative proportions as given below:—

DIAMETER.				CLAY SUBSOIL. PER CENT.	LOESS SOIL. PER CENT.
3937 to 4724 hundred thousandths of an inch				.8	.2
2362 to 3937	"	"	"	1.2	
1575	"	"	"	2.0	.4
1181	"	"	"	1.6	.6
610	"	"	"	.9	.9
472	"	"	"	.3	1.7
283	"	"	"	.2	2.0
185	"	"	"	2.5	14.3
142	"	"	"	3.7	16.2
98	"	"	"	5.6	20.1
59	"	"	"	10.6	5.6
31	"	"	"	24.7	33.6
4	"	"	"	48.0	2.5

The values given in these tables have great significance in showing, first, how it is that soils can act as water reservoirs of great capacity; second, how even when they are made up very largely of materials extremely difficult of solution, they may, when kept supplied with an abundance of water, dissolve in large quantities; and third, how it is that the roots of plants are brought into very intimate contact with an immense surface of soil grains.

If you plunge a marble into water, and then withdraw it, it comes forth surrounded by a film and the greater the surface of the marble the larger will be the amount of water which adheres to it. When the rains saturate any soil, the surface of each soil grain has its film of water as in the case of the marble. It is not difficult to see how reducing the diameters of soil grains, greatly increases the number of them in a cubic inch, and at the same time increases the total soil surface to which the soil water may adhere. Suppose we take a marble exactly one inch in diameter. It will just slip inside a cube one inch on a side and will hold a film of water 3.1416 square inches in area. But reduce the diameters of the marbles to one-tenth of an inch and at least 1000 of them will be required to fill the cubic inch, and their aggregate surface area will be 31.416 square inches. If, however, the diameters of these spheres be reduced to one-hundredth of an inch, then 1,000,000 of them will be required to make a cubic inch, and their total surface area will then be 314.16 square inches. Suppose again the soil particles to have a diameter of one-thousandth of an inch. It will then require 1,000,000,000 of them to completely fill the cubic inch, while their aggregate surface area must measure 3141.59 square inches.

Turning back to the tables which show the observed sizes of grains obtained by the mechanical analyses, it will be noted that the smallest of them have diameters placed at .00004 of an inch, a size far below the last we have considered, and yet a cubic foot of soil grains having a diameter of .001 of an inch will provide a surface for holding water, as we have seen the marble do, equal to 37,700 square feet. Four feet in depth of such a soil, a depth to which the roots of most farm plants penetrate, would possess a water-holding surface not less than 3.4 acres for each column of soil one square foot in section. An extremely thin film, therefore, on such a surface, must aggregate a large amount of water, and we need not be surprised to learn that only moderately fine-grained soils have been observed to retain in the surface five feet, under field conditions, no less than the equivalent of 12 to 20 inches of water on the level.

It should be observed here, regarding the very small size of grains shown in the tables of mechanical analyses given above, that it is extremely improbable that the smallest of them are individually invested by films of water as these occur undisturbed in the soil. On the contrary, future study seems likely to demonstrate that even the stiffest clay soils are made up of complex grains, into which the capillary waters do not freely pass, and that the extreme division shown by present methods of mechanical analysis are in part the result of a breaking down of compound grains which, under field conditions, act largely as if they were solid particles of greater size. These statements are made here to caution the reader against carrying the computations made above by the writer much farther than he has done.

Turning now to the influence of the texture of the soil on its ability to supply the soluble ash ingredients needed by the vegetation growing upon it, it will be at once apparent that the broad surface which the soil water has been shown to cover must lead to a rapid solution of soil material and a relatively rapid production of plant foods, when they are present, even though they may exist in the soil in forms soluble with extreme difficulty.

An instructive experiment has been made by the French chemist Pelouze, aiming to illustrate the influence of a fine state of division, or of small particles, on the rate of solution. In a flask holding a little more than a pint, he kept water boiling constantly during five days, and at the end of this time he emptied, dried, and weighed the flask, finding it to have lost less than two grains in weight. He next broke off the neck of the flask; grinding it to a fine powder and returning it to the body of the flask, he repeated the boiling during five more days. At the end of this time he found that fully one-third of the total weight of his flask had been dissolved by the water.

Storer states that the Messrs. Rogers found, by digesting powdered feldspar, hornblende, and various other minerals with water for a week, that from a third of one per cent to one per cent of the mineral was dissolved out by the water. Taking the lowest estimate, namely one-third of a pound of mineral for each one hundred pounds of the dry powdered rock, and multiplying the result by the weight of soil in the surface foot of an acre of land, we find that more than 10,000 pounds might be so dissolved during one week. But Voelcker has shown that 15 tons of good well-rotted manure to the acre will supply no more than 150 pounds of potash and 140 pounds of phosphoric acid to the acre.

It is undoubtedly true that the rate of solution here cited is much more rapid than ever occurs in natural soils, but the work may go forward more or less rapidly during the whole year and throughout a depth of more than five feet, and still be within the reach of root action; and more than this, it is probable that the state of division in the soil much exceeds that which obtained in the experiment of the Messrs. Rogers. It is not strange therefore that, where due attention has been paid to rotation of crops and where a fair amount of organic matter in the form of stable manure has been supplied, many soils have been found practically inexhaustible even under centuries of cropping.

When we bear in mind that the roots of our cultivated plants penetrate the soil to a depth of four or more feet, that they extend horizontally away from the stems through an equally great distance, and that throughout all of this broad and deep mass of soil innumerable and extremely minute root hairs thread their way among the soil grains and come in touch with them throughout their whole length and on all sides, we are confronted with the great work which is conjointly done by the sunshine, the plant, and the soil. Viewed in this light, soil is not a grave where death and quiet reign, but rather a birthplace where the cycles of life begin anew, to run their courses over and over again.

There are some soils which, in their minuter structure, are made up of extremely small particles, but which are nevertheless so coarse-grained as to be more or less leachy in character and to possess a quality of easy tillage approaching that of sandy land. The *loess* soils are a type of these, and they contain a large number of compound grains in which the minute particles have been cemented with carbonates of lime and magnesia into small concretions which behave, so far as the texture of the soil is concerned, as though they were in reality grains of sand. There must also be some of this sort of structure in the more open types of clay soils, and it is probable that the underdraining of such lands tends to augment this tendency to granulation and thus improve their quality. Stiff, clayey subsoils, too, have a tendency to shrink and draw together into small cube-like blocks, particularly during dry seasons and especially after they have been underdrained, and this structure facilitates in a marked manner complete drainage, thorough aëration and a deeper penetration of the roots of plants into them. Such an opening up of a clay soil as this makes it possible for the carbonic acid of soil water to dissolve lime and bring it into contact with the deeper and stiffer clays, and, by its flocculating or granulating tendency, to so improve their texture, in the course of years of proper tillage, as to render them more productive and more easily worked.

CHEMICAL CONSTITUENTS OF SOILS.

From what has been said regarding the origin of soils, it will be evident, at once, that whether we look at them from the standpoint of the chemical elements which enter into their composition, whether we enumerate the chemical compounds which a thorough analysis might reveal, or whether we attempt to name the various minerals which time and the agents of geologic change have brought together, we must expect to find almost any soil a body of extreme complexity, so far as its composition is concerned. The fact that few indeed, if any, of the minerals entering into the rock structure of land masses are wholly insoluble, even in pure water, the fact that this water carries its dissolved materials both through and across the surface of the soil and over long distances, and the fact that the movements of the surface waters, winds, and burrowing animals of all kinds and even plants work together in mixing and transporting soil particles, must tend to give to all soils, wherever they may be found, a general composition which is extremely similar.

Of the many chemical elements occurring in the soil, those which have the greatest abundance, or are of most importance from the standpoint of agriculture, are named below: —

Non-Metals.		Metals.	
Oxygen.	Chlorine.	Aluminum.	Sodium.
Silicon.	Phosphorus.	Calcium.	Iron.
Carbon.	Nitrogen.	Magnesium.	Manganese.
Sulfur.	Fluorine.	Potassium.	
Hydrogen.	Boron.		

Oxygen occurs in the soil in the free state and in combination with all of the elements named except fluorine. Silicon, combined with oxygen, forms silica or quartz, so familiar to us in the form of plastering sand, pebbles of quartz, carnelian, and jasper. So abundant is silica or quartz that it is estimated to compose more than one-half of the known rocks of the earth. Because it is so difficultly soluble and so hard, quartz, in larger or smaller particles, is the chief ingredient by volume and by weight of nearly all soils everywhere.

Carbon occurs in the soil as a part of the humus or organic matter, united with calcium and with magnesium in the form of carbonates, and also united with oxygen in the form of carbon dioxide as one of the soil gases and which, dissolved in the soil water, plays so important a part in the solution of plant food. But the plant gets the carbon of its food from the free carbon dioxide of the air.

Sulfur occurs in the soil in the form of sulfids united with iron or as sulfates united with some metal, as iron, lime, or magnesia. United with oxygen and calcium, it forms land plaster or gypsum, an important fertilizer; it is also an essential part of many organic compounds in the tissues of plants and animals.

Hydrogen plays its greatest part in agriculture, and in the life of the world, while combined with oxygen in the form of water, and it is in the form of water that plants get most of the hydrogen and oxygen which go to make up their tissues and the starches, sugars, and other forms of stored food.

Chlorine is neither a large constituent of soil nor one which plays a very important part in its work. It is usually associated with sodium in the form of common salt, which occurs in all soils, in all natural waters, and even in rain. Chlorine is uniformly found in the composition of plants, and is regarded as in some way essential.

Phosphorus, which in its uncombined state is used in the manufacture of matches, because it so readily ignites in the presence of oxygen, is never found in the soil or in nature except combined with some other substance. It is very generally distributed in the soil, though not in large quantities, but is a very essential soil ingredient. It occurs as a constituent of the oldest known rocks and from these, through the processes of rock decay and rock building, it has been ever present in the soils of all geologic ages. Through the action of plants it is gathered up from the soil and concentrated in their tissues; animals in their turn, feeding upon the plants, or upon other animals, concentrate the compounds of phosphorus still farther, so that now the remains of animal life in rock deposits are valuable sources of commercial fertilizers, and man, through his intelligent effort, is taking advantage of these ancient concentrations of plant food by bringing them forth and redistributing them where they may be used over again in carrying forward once more the life processes of the present day.

Nitrogen, though the most abundant element in the atmosphere, is one of the least abundant in the crust of the earth as a rock ingredient. In the soil, in the combined form, it occurs as a part of humus and the fragments of the decaying tissues of plants and animals, from which, through the instrumentality of microscopic life there, it is converted ultimately into nitric acid, which, uniting with potash, lime, or other soil ingredients, forms a soluble salt taken up by the roots of plants and is then made to yield up its nitrogen to build those nitrogenous compounds so abundant in the tissues of animals.

Neither boron nor fluorine have any considerable abundance in soils, and their percentage composition in the ash of plants is small. Borax is the most familiar compound in which boron occurs. Fluorine, united with lime, occurs in Iceland spar, and it is a constituent of the blood, milk, teeth, and bones of mammals.

Aluminum is placed third in abundance among the elements of the surface ten miles of the solid land. It is a beautiful white metal, remarkably light, and does not readily tarnish. While it is not a plant food, it is a fundamental constituent of true clay, which is derived from the breaking-down, by chemical action, of feldspar, mica, and other constituents of granitic rock; and it is important to keep in mind that the true clays, and the fine clay-like particles play a very important part in determining the texture and water-holding power of soils.

Calcium and magnesium are two metals very closely associated, and, united with carbon dioxide, they constitute the limestone beds of the world. Magnesium is an indispensable plant food, as is also lime. Both collect largely in the seeds of plants, the magnesia more abundantly than the lime, wheat containing 12 per cent in its ash of magnesia, as against 3 per cent of lime, while the ashes of peas contain 8 per cent of the former to 4 per cent of the latter. If phosphorus is counted a very important plant food because it occurs in the seeds of plants, the same reasoning would be applicable to both lime and magnesia, but both of these substances occur in most soils in much larger quantities than does phosphorus, and because of this their need is not so often felt.

Potassium is one of the very essential elements of plant food, and although it is almost universally distributed in soils, the compounds which it forms are usually so soluble that there is a strong tendency for them to leach away and to be borne to the sea in the drainage waters. On account of this characteristic of the potash salts, bad management leads to a rapid depletion of the soil stores of this element and a corresponding shortage in the yields of crops. Potash occurs widely distributed in the earth's crust, largely as a constituent of orthoclase feldspar and the biotite and muscovite micas. In certain places, too, it occurs in the mineral glauconite of the greensands of New Jersey and elsewhere, and also in certain layers of the potsdam sandstone in various parts of the world. The glauconite sometimes composes as much as 90 per cent of the greensands and is itself made up of 8 to 12 per cent of potash. Potash, too, is also a notable constituent of kaoline beds, those in Wisconsin showing an average per cent of 1.975 for fifteen analyses.

Repeatedly, in geologic history, movements of the earth's crust have shut off arms of the sea, whose waters afterward evaporated, leaving very rich and extensive deposits of those salts which were carried in solution by the ocean waters; and among these were laid down vast beds of kainite which are now being mined and returned as fertilizers to the land, where the potash is used over again. The ashes of land plants are rich in potash as a carbonate, and, united with nitric acid, it is produced in the soil in the form of nitre.

Sodium is the basis of common salt, and as such has a world-wide distribution. It very much resembles potassium as an element but can in no sense take its place in the life of land plants. In the form of Chili saltpetre sodium nitrate is largely used as a fertilizer, but for the nitric acid it contains rather than for the sodium.

Iron and manganese are two of the most abundant of the heavy metals and occur almost universally as constituents of the soil, but the former in much larger quantities than the latter. Iron, united with oxygen and with water, constitutes the red and yellow ochres used so much as pigments in painting, and which give to soils their red and yellow colors. Iron, too, is an important plant food, although it does not enter largely into the composition of their tissues. It is so abundant, so universely present in the soil, and so difficultly soluble that there is never a deficiency of it so far as the purposes of the plant are concerned.

If we look at the composition of soils as shown by chemical analysis, it will be observed that while there is a strong likeness among them all, to which reference has been made, there are nevertheless quite wide variations among them, so far as the relative proportions of the constituents are concerned; and the truth of this statement will be readily appreciated from the tables on pages 84–87, which are taken from the papers of Hilgard in the tenth census of the United States. Selecting for comparison the most widely divergent types, so far as physical conditions are concerned, we have set against the sandy soils the heavy clays.

That these analyses may have as great significance as possible, brief descriptions of the soils are given in the order in which they occur in the table, and the sandy soils are first described under the numbers used by Hilgard.

No. 36. Little Mountain soil, Alabama. Native vegetation, chestnut, short-leaf pine, hickory, post oak, and small sourwood. Depth 8 inches; color, top soil, dark brown 2 inches; yellowish sand at 2 feet and at 5 feet solid sandstone rock.

No. 11. Chunnenugga ridge soil, Alabama. Depth, 6 inches. Vegetation, chestnut, short-leaf pine, red oak, and sour gum; color, dark gray, changing at 6 inches to a lighter gray, and at 3 feet to a yellowish color.

No. 8. Pine upland soil, table-land, Florida. Vegetation, long-leaf pine, round and narrow leaf, black-jack, red and post oaks, and hickory. Soil taken at a depth of 9 inches. Subsoil, a yellowish sand, with slight intermixture of yellowish clay, becoming a hard yellow clay at a depth of 2 to 6 feet.

No. 7. Gray, sandy pine woods soil, Florida. Depth, 10 inches. Vegetation, long-leaf pine and wire grass.

No. 142. Gray, sandy soil, Georgia. Taken 6 inches deep. Vegetation, red and post oak, pine and hickory.

No. 322. Dark, sandy, upland soil, Georgia. Taken 6 inches deep. Timber growth, hickory, oak, and long-leaf pine.

No. 359. Gray, sandy soil, Georgia. Depth about 6 inches, with yellow sandy subsoil.

No. 252. Dark, sandy soil, Georgia. Depth, 6 inches, underlaid at a depth of a few feet by white marl beds. This and No. 359 are described as oak, hickory, and pine uplands.

No. 307. Gray, sandy soil of oak and hickory lands, Georgia. Sarsaparilla in abundance. Depth, 6 inches.

No. 165. Gray, sandy soil, Georgia. Depth, 6 inches. Vegetation, long-leaf pine.

The clay soils are as follows: —

No. 98. Post-oak and flatwoods clay, Alabama. Depth, 10 inches, reddish clay, spotted. Vegetation, chiefly post oak.

No. 149. Red clay soil, Georgia. Vegetation, red, white, and post oaks, dogwood, chestnut, hickory, and pine. Depth, 5 inches.

No. 203. Deep red soil, Georgia. Growth, red and post oak and hickory. Depth, 8 inches.

No. 166. Red hill lands, Georgia. Taken 6 inches deep. Vegetation not given.

No. 513. Ash-colored clayey swamp land, Georgia. Growth, cypress, water oak, gum, ash, maple, beech, and saw palmetto. Blue clay stratum at 1 to 6 feet.

No. 10. Yellowish red clay, South Carolina. Timber, post, white, and black oaks, short-leaf pine, and hickory. Depth, 5 inches.

No. 141. Stiff red soil from the cretaceous prairie

CHEMICAL COMPOSITION OF SOILS.

Numbers.		Insoluble Residue.		Soluble Silica.		Total Insoluble Residue and Soluble Silica.	
Sand.	Clay.	Sand.	Clay.	Sand.	Clay.	Sand.	Clay.
36	98	93.630	72.746	1.682	8.926	95.312	81.672
11	149	94.770	73.690	.486	3.370	95.256	77.060
8	203	93.362	60.370	1.721	2.000	95.083	62.370
7	166	95.630	73.422	.879	2.709	96.509	76.131
142	513	92.090	63.444	1.220	11.325	93.310	74.769
322	10	90.230	77.860	1.940	1.790	92.170	79.650
359	141	90.681	54.565	1.885	13.219	92.566	67.784
252	390	92.460	51.063	1.550	20.704	94.010	71.767
307	1	94.428	79.580	.529	3.628	94.957	83.208
165	7	94.822	75.350	1.037	7.310	95.859	82.660
Averages.........		93.210	68.209	1.293	7.498	94.503	75.707
	374		91.498		1.722		93.220

SWAMP AND LOESS SOILS.

	Humus.	Loess.	Humus.	Loess.	Humus.	Loess.
Averages.........	35.886	68.853	20.825	4.918	56.711	73.771

SOILS COMPARED WITH THEIR SUBSOILS.
SOILS.

	Sand.	Clay.	Sand.	Clay.	Sand.	Clay.
Averages.........	93.222	73.978	1.019	5.034	94.241	79.012

SUBSOILS.

Averages.........	90.714	66.290	2.212	7.446	92.926	73.736
Differences.......	+2.508	+7.688	−1.193	−2.412	+1.315	+5.276

ARID AND HUMID SOILS COMPARED.

	Humid.	Arid.	Humid.	Arid.	Humid.	Arid.
Averages.........	84.031	70.565	4.212	7.266	87.687	76.135

CHEMICAL COMPOSITION OF SOILS.—Continued.

Numbers.		Potash.		Soda.		Lime.		Magnesia.	
Sand.	Clay.	Sand.	Clay.	Sand.	Clay.	Sand.	Clay.	Sand.	Clay.
36	98	.100	.416	.060	.112	.120	.080	.040	.691
11	149	.156	.176	.069	.004	.081	.090	.069	.112
8	203	.045	.186	.018	.119	.064	.071	.005	.065
7	166	.117	.134	.064	trace	.058	.219	.042	.289
142	513	.110	.242	.035	.079	.090	.387	.025	.508
322	10	.067	.092	.009	.041	.119	.036	.090	.070
359	141	.275	.431	.130	.277	.055	.540	.048	.836
252	390	.095	1.104	.036	.325	.076	1.349	.083	1.665
307	1	.209	.150	.069	.065	.141	3.054	.031	.029
165	7	.034	.255	.022	.258	.045	.340	.043	.296
Averages.....		.121	.319	.051	.128	.085	.617	.048	.456
	374		.137		.054		.173		.203

SWAMP AND LOESS SOILS.

	Humus.	Loess.	Humus.	Loess.	Humus.	Loess.	Humus.	Loess.
Averages.....	.639	.435	.109	.165	3.786	5.820	.886	3.692

SOILS COMPARED WITH THEIR SUBSOILS.
SOILS.

	Sand.	Clay.	Sand.	Clay.	Sand.	Clay.	Sand.	Clay.
Averages.....	.157	.214	.072	5.08	.115	1.761	.076	.182

SUBSOILS.

Averages.....	.143	.344	.064	.085	.096	1.481	.073	.240
Differences...	+.014	−.130	+.008	.000	+.019	+.280	+.003	−.058

ARID AND HUMID SOILS COMPARED.

	Humid.	Arid.	Humid.	Arid.	Humid.	Arid.	Humid.	Arid.
Averages.....	.216	.729	.091	.264	.108	1.362	.225	1.411

CHEMICAL COMPOSITION OF SOILS. — Continued.

Numbers.		Brown Oxide of Manganese.		Peroxide of Iron.		Aluminum.	
Sand.	Clay.	Sand.	Clay.	Sand.	Clay.	Sand.	Clay.
36	98	.102	.106	.761	12.406	1.532	2.473
11	149	.156	.146	.706	5.989	.733	7.305
8	203	.220	.196	.941	9.709	1.339	18.066
7	166	.049	.164	.224	4.054	.473	10.598
142	513	.126	.052	.963	3.894	1.959	13.454
322	10	.313	.056	1.927	5.646	2.141	7.538
359	141	.172	.079	1.837	7.089	1.436	16.071
252	390	.040	.119	.843	5.818	2.649	10.539
307	1	.101	.195	.661	3.420	1.195	4.988
165	7	.020	.038	.930	5.784	1.576	5.567
Averages.......		.130	.115	.979	6.381	1.503	9.660
	374		.066		1.372		1.522

SWAMP AND LOESS SOILS.

	Humus.	Loess.	Humus.	Loess.	Humus.	Loess.
Averages.......	.098	.164	7.040	3.569	14.476	2.812

SOILS COMPARED WITH THEIR SUBSOILS.
SOILS.

	Sand.	Clay.	Sand.	Clay.	Sand.	Clay.
Averages.......	.124	.133	1.162	5.205	1.145	6.998

SUBSOILS.

Averages.......	.080	.125	1.739	6.947	2.276	12.086
Differences.....	+.044	+.008	—.577	—1.742	—1.131	—5.088

ARID AND HUMID SOILS COMPARED.

	Humid.	Arid.	Humid.	Arid.	Humid.	Arid.
Averages.......	.133	.059	3.131	5.752	4.296	7.888

Chemical Composition of Soils.

CHEMICAL COMPOSITION OF SOILS.—Concluded.

Numbers.		Phosphoric Acid.		Sulfuric Acid.		Water and Organic Matter.	
Sand.	Clay.	Sand.	Clay.	Sand.	Clay.	Sand.	Clay.
36	98	.051	.103	.028	.061	2.055	1.906
11	149	.101	.071	.057	.055	2.642	8.891
8	203	.066	.204	.091	.285	2.422	8.953
7	166	.092	.069	.058	.035	1.807	8.309
142	513	.191	.071	.105	.055	3.477	6.843
322	10	.111	.082	.054	.054	2.881	6.167
359	141	.105	.187	.034	.009	3.682	6.922
252	390	.039	.304	.045	.024	2.354	7.369
307	1	.103	.242	.046	.089	3.113	4.962
165	7	.014	.079	.035	.079	1.636	4.962
Averages......		.087	.141	.055	.075	2.607	6.528
	.374		.088				3.394

SWAMP AND LOESS SOILS.

	Humus.	Loess.	Humus.	Loess.	Humus.	Loess.
Averages......	.150	.200	.148	.090	13.943	1.205

SOILS COMPARED WITH THEIR SUBSOILS.
SOILS.

	Sand.	Clay.	Sand.	Clay.	Sand.	Clay.
Averages......	.128	.207	.052	.090	2.853	6.014

SUBSOILS.

Averages......	.124	.159	.060	.071	1.943	4.780
Differences....	+.004	+.048	—.008	+.019	+.910	+1.234

ARID AND HUMID SOILS COMPARED.

	Humid.	Arid.	Humid.	Arid.	Humid.	Arid.
Averages......	.113	.117	.052	.041	3.644	4.945

region, Mississippi. Fairly productive in good seasons. Vegetation, oak.

No. 390. Buckshot soil of Yazoo bottom, Louisiana. Stiff dark-colored clay soil, mottled with spots of ferruginous matter, traversed by numerous cracks. Very fertile and no change in character to a depth of 10 feet. Vegetation, sweet gum, pecan, water and willow oak, hackberry, and honey locust.

No. 1. Red clay land from central basin, Tennessee. Vegetation, hickory, red, white, and post oaks, elm, ash, honey locust, black walnut, wild cherry, sugar trees, poplar, hackberry, redbud, dogwood, and papaw. Depth, 7 inches, with heavy clay subsoil.

No. 7. Same region, depth, and vegetation as No. 1, but subsoil not as heavy.

If reference is made to the table showing the chemical composition of sandy soils as compared with that of clay soils, it will be seen that during the process of analysis 31.791 per cent of the soil ingredients were dissolved out of the clay, as shown by deducting the 68.209 per cent of insoluble residue from 100, while only 6.790 per cent were dissolved from the sandy soils, making a difference of 25.001 in favor of the clay soil. In the average, too, of the ten different analyses, it will be seen that the various soluble ingredients are shown to be dissolved more abundantly from the clayey than from the sandy soil.

How shall these results be understood? Do they mean that in every 100 pounds of sandy soil, plants may find nearly 7 pounds of soluble material, the larger part of which is plant food, while in the clayey soil there are 31 pounds in each and every 100, or more than four times as much? Do they mean that clay soils are capable of yielding the ash ingredients of plant food to growing crops four times more rapidly than sandy soils can do? Or, are we to understand that the ash ingredients of plant food in clay soil are being carried away by the waters which percolate through them four times more rapidly, gallon for gallon of water, than they are from the sandy soil?

If the results of the chemical analyses of soil were to be taken without qualification, one or the other of these conclusions would seem to follow, but it is very unfortunate for agriculture that it should seem necessary to admit that the results of soil analyses, as they have been made, can and do throw but a very dim and uncertain light upon either the condition or the amount of available plant food a soil may contain. The results of chemical analyses do show, beyond question, that there are marked differences between soils, but so far as the data now at hand can be interpreted, these variations, at least so far as the mere relative proportions of the so-called soil ingredients are concerned, seem quite as likely to be due to differences in the actual sizes of soil grains, as to any real and marked difference in either the chemical or mineralogical composition of them.

When given quantities of coarse and fine grained soils, having identical chemical composition, are subjected to like quantities of dissolving acid, under like conditions, it would appear to follow necessarily that, where all of the materials are not dissolved, and where a complete saturation of the solvent has not resulted, more materials will enter into solution from the fine than from the coarse grained sample; and simply because more points of attack are presented by one than by the other. With like quantities of soil and acid, the solution would not be directly proportional to the total surface areas in the two samples, because in the fine-grained sample, where the solution is most rapid on account of the greater surface, the strength of the acid would be most rapidly reduced, thus tending to make the ratio between the quantities dissolved somewhat less than the ratio between the areas of the soil grains in the two samples.

If it should occur that certain kinds of soil grains in both cases were more easily and completely dissolved than the bulk of the soil materials, then the ratio, in such cases, between the dissolved materials would be less than between the surfaces of the original soil grains. So, too, unless very vigorous stirring were constantly maintained, the fine-grained soil would tend to dissolve less rapidly in proportion to its surface than would the coarse-grained soil, where diffusion would be relatively more rapid.

But in spite of the condition just referred to, the much larger surface area possessed by the two, three, or more grams of clay soil taken for analysis appears to have resulted in extracting a relatively larger amount of plant food from this soil than was removed from the more coarse-grained and relatively small-surfaced sandy soil.

If we compare the mechanical analyses of the two Mississippi soils, Nos. 390 and 377, as given by Wiley, with their chemical analyses, as given by Hilgard, it will be found that, while the surface area in one gram of the fine soil is about seven times that of the one of coarser grain, the soluble materials extracted were only about four times as large, thus making the rate of solution, when compared by equal surface areas, less for the clay than for the sand. So, too, if we take the mean surface area in one gram of twelve truck subsoils, of Maryland, as given by Whitney, and compare this with the mean area of one gram of six strong wheat subsoils, we shall find their surface areas to be in the ratio of 934 to 3451, while the ratio between the soluble ingredients of the sandy and clayey subsoils, as given on pages 84–87, is 9.286 to 33.71, or very nearly the same as that between the surface areas of the soil grains in the similar cases cited.

The reader should bear in mind that these remarks are not intended to convey the idea that the chemical composition of sandy and clayey soils are nearly identical, but rather that it seems more than probable that the ratio between the more difficultly soluble and the more readily soluble ingredients of plant food is not as widely different in the coarse and fine grained types of soil as the table of chemical analyses would appear at first thought to indicate.

There has been placed in the table, just below the line of averages for soils, the analysis of a clay soil, No. 374, from Louisiana, which shows a relative proportion of the more soluble ingredients bearing a striking resemblance to the more sandy types, the insoluble residue being 91.498 per cent, while that of the sandy soil averages 93.210, there being two samples in the series of ten having as low a per cent of "insoluble residue" as 90.6. Hilgard describes this as a fair upland soil, yielding 700 to 800 pounds of seed cotton per acre, color gray, depth 6 to 8 inches, not heavy, and underlaid by a subsoil quite heavy in tillage and dark orange in color. The case is cited to illustrate how a chemical analysis alone does not always serve to distinguish between soils which in their physical features may be strongly contrasted.

If we look at the differences between the indicated

chemical composition of soils and their subsoils, as given in the table, pages 84–87, it will be seen that the insoluble residue, lime, phosphoric acid, and manganese are more abundant in the surface soil, while in the subsoils the soluble silica, peroxide of iron, and alumina appear in highest percentages, while the other ingredients are sometimes more abundant in one and sometimes in the other.

If these soils were examined with reference to the sizes of the grains composing them, it would be found that in the subsoil the small particles, especially in humid regions, are most abundant, and hence that the internal surface area is much larger in the lower than in the upper layers, which may in part explain the higher percentages of soluble ingredients shown by chemical analysis.

Referring again to the table for a comparison of soils of humid with those of arid regions, it will be observed that these are in some respects quite strongly contrasted as regards their chemical composition. The averages there given are those of Hilgard, and for the humid regions are made up of 466 analyses, while those for the arid are a mean of 313 determinations.

It may be noted in the first place that, excepting the insoluble residue, brown oxide of manganese and sulfuric acid, all other ingredients are more abundant in the soils of arid lands; and not only this, but these differences are really large, the amounts being often more than double, and in some cases much higher than this. It should be said, however, regarding the analyses entering into these averages, that all of them come from states south of the Ohio River or west of the Missouri, and for this reason more fairly represent differences between soils of unglaciated areas in humid and arid regions than of soils in general. It appears, too, that many of the differences between the soils of the two regions largely represented are made more striking than they should be through the admission of a relatively larger proportion of sandy soils and light loams than of those of heavier type, and this will be rendered apparent on comparing the soils of the arid regions with the ten heavy clay soils in the same table, and still more so when the clay subsoils are compared with them. The position would seem to be more fully justified, too, when it is observed that the analyses of the ten sandy soils given in the table approach more nearly to the mean of the 466 for humid regions than do those of the clay soils; and also when it is observed that the brown oxide of manganese, which is more abundant in humid than in arid soils, appears also to be more abundant in the sandy than in the clay soils.

But when all allowances have been made, there are still outstanding differences between the soils, formed under these widely contrasted climatic conditions, which are sharply marked and beyond question.

In the first place, the insoluble residue is evidently more abundant in humid than in arid soils, the ratios being in round numbers, as 84 to 70, and this difference is held to be due to the much greater amount of leaching which necessarily takes place where rains fall frequently and in large quantities.

The soluble silica, too, an ingredient characteristic of the analyses of only heavy soils in humid regions, is very abundant in those of arid regions, but the anomalous feature regarding this is that in its combinations in the soil it has not the power of imparting to these soils the adhesive plastic quality so characteristic of clay lands in humid regions. Whether the soluble silica is, in the soil, united largely with alumina to form kaolinite or true clay, or whether it is combined to form varieties of zeolites, which may be lacking in power to give adhesiveness to soil, is unknown.

Hilgard regards the large percentage of lime, with its allied compound, magnesia, as one of the most distinctive features of the soil of arid regions, and to this lime he attributes a flocculation or granulation of the clay, which destroys its adhesive quality. To the very general abundance of lime, too, in these soils, alike on the high lands and on the low, he attributes the high and very uniform productiveness of all these soils whenever an abundance of water is supplied to them.

It seems plain that the high percentages of soluble ingredients in the soil of arid regions result from the slow decomposition of soil grains brought about by the conjoint action of the scanty water which does fall and the carbonic acid of the air. Water enough falls for the decomposition of rock and the formation of alkalies and zeolitic minerals, but not enough to remove them when formed, as is the case in humid regions.

The humus of soils, so far as its chemical composition is concerned, is not well understood, neither have we attained a satisfactory knowledge of its functions or importance as a food for plants. It used to be held that any soil deficient in humus was, because of this shortage, necessarily poor or sterile, but it is now known that in arid regions, where humus in the soil is very scanty or even wanting, large crops are produced when only an abundance of water is supplied. Experiments in water culture, too, have proved that when nitrogen is supplied to plants in the form of purely inorganic or mineral nitrates, plants will thrive in the complete absence of humus.

Humus is derived from the partial decay in the soil of organic matter, whether that be vegetable or animal, and it imparts to the soil a black or brown color. From what has been said regarding nature's method of running in cycles, it will be readily understood that humus is simply the abandoned tissues out of which life has moved and which are falling back by degrees into the carbonic acid, water, free nitrogen, and ashes out of which life reared its marvelous structures. In tropical regions, where the soil is warm the whole year through, and in arid regions, where the soil is open and readily penetrated by the air, the rate of decay is so rapid that the amount of humus found in any soil at any particular time is relatively small; but in the temperate climates, where the soil is damp and where the ground is too cold to permit of decay during considerable portions of the year, there the decaying organic matter accumulates in considerable quantities and especially on wet soils and in swampy places. Beds of peat and the black muck soils are the best examples of what is meant by humus, and it is excessively abundant in these places because the soil is close in texture and so full of water that the microscopic forms of life which feed upon this dead matter are unable to get the necessary oxygen to thrive in it. We should think of humus as the food of microscopic life in the soil, and of the waste products of this microscopic life as a very essential part of the food of higher plants. Keeping this in mind, we can better appreciate the importance of farmyard manure, for through its decay humus is formed. Both are organic matter partially decayed and capable of contributing food to plants.

But Hilgard and Jaffa have made the important announcement that the humus of arid regions is much richer in nitrogen than is that of humid regions, or of the humid soils of arid regions. The results of their observations, as reported in *Agricultural Science* for 1894, are summarized below: —

	No. Samples.	Humus in Soil. Per Cent.	Nitrogen in Humus. Per Cent.	Humic Nitrogen in Soil. Per Cent.
Arid soils	18	.75	15.87	.101
Semi-arid soils	8	.99	10.03	.102
Humid soils	8	3.04	5.24	.132

In speaking of these results, they say, "It thus appears that, on the average, the humus of the arid soils contains three times as much nitrogen as that of the humid, that in the extreme cases the nitrogen percentage in the arid humus actually exceeds that of the albuminoid group, the flesh-forming substances."

"It thus becomes intelligible that in the arid region a humus percentage, which, under humid conditions, would justly be considered entirely inadequate for the success of normal crops, may, nevertheless, suffice even for the more exacting crops. This is more closely seen on inspection of the figures in the third column, which represent the product resulting from the multiplication of the humus percentage of the soil into the nitrogen of the humus."

FUNCTIONS OF THE SOIL INGREDIENTS.

When we come to speak of the functions or importance of the different soil ingredients in vegetable life, it must be said that our knowledge is as yet very limited and very indefinite. That plants cannot thrive in a soil destitute of nitrogen, potash, lime, magnesia, and phosphoric acid has been proved in the most complete and satisfactory manner. A soil entirely lacking in any one of these is, for that reason, an infertile one.

Since sulfuric acid, in some of its combinations in the soil, is the only known source of sulfur in plants, and since sulfur is an essential element in the molecular constitution of vegetable albumen and allied compounds, it follows that fertile soils should always contain an adequate percentage of sulfates in some form available to plants. It will be seen, however, from the table of chemical composition, that the amount present in any soil, at any time, is usually relatively small.

Silica, although the most abundant of all the soil ingredients, and although present in the tissues of nearly all plants, in greater or less amounts, has not been demonstrated to have any important part to play in plant growth. Indeed, recorded observations go to show that plants get along perfectly well when grown in media nearly or entirely devoid of this substance. It has been supposed that it played an important part in giving the necessary stiffness to the stems of cereals, but Arendt has shown that the distribution of silica in different parts of the oat plant is as follows: —

Lower part of stem	1.7	parts in 1000 of dry substance.
Middle " " "	5.1	" " 1000 " " "
Upper " " "	13.3	" " 1000 " " "
Lower leaves	35.2	" " 1000 " " "
Upper leaves	43.8	" " 1000 " " "

Now if the silica were needed for stiffness, the largest amounts should be found in the lower and middle portions of the stem, and more in the stalk than in the leaves, while the reverse distribution is observed. Indeed, the distribution of silica in the oat makes it appear that its presence there is due simply to the fact that it was present in the water which passed through the plant, and that it was left where the largest amounts of evaporation had taken place; that is, in the leaves and in the higher portions of the stem where the wind currents are strongest.

A sufficient amount of an iron salt in the sap of a plant appears to be essential to the development of the green color of foliage, and without the green chlorophyll carbon cannot be assimilated from the carbon dioxide of the air. Sachs found, that while young plants of Indian corn, when growing in solutions free from iron, had their first three or four leaves green as usual, the next, and those which followed, were first white at the base and green at the tips and afterwards completely white; but on adding a few drops of sulfate or chloride of iron to the medium in which the plants were growing, the green color began to be developed appreciably in twenty-four hours, while the normal color was restored in three or four days. It is plain, therefore, that iron is a very important soil ingredient, although its universal presence in the soil, and the small amount of it needed, makes it unnecessary to pay any heed to it as a plant food which should be supplied as a fertilizer under natural field conditions.

In the movements which result in the transfer of the starch-forming products from the green parts of plants to their place in the seed, root, or tuber, potash, magnesia, and lime appear to play an important part, but it seems also that before the starch-forming products can be moved a small amount of chlorine must also be present. The amount of chlorine, however, may be very small, and as it does not tend to accumulate in the places where the starch is stored, it would appear that, like the iron, it may be used over and over again.

KINDS OF SOILS.

In the language of practical men soils are variously classified. They are sometimes spoken of as light or heavy, but these terms do not refer to their weight in pounds per cubic foot; for those which are called the lightest are the heaviest soils we have. A heavy clay soil, when dry, weighs only from 70 to 80 pounds per cubic foot, while the lightest sandy soils have a weight of 105 to 110 pounds for the same volume. It is the ease or the difficulty with which the soils are worked or tilled which gives rise to the terms, light and heavy. In the light soils the roots of plants have less difficulty in threading their way than in those which are stiff and not easily crowded to one side, and as a result of this freedom to movement plants distribute their roots much more symmetrically and occupy the whole soil more completely in the lighter types than they are able to do in the stiff and heavy ones. As a consequence of this equitable distribution, no root trespasses upon the feeding ground of another, and the whole soil is laid under tribute with correspondingly better returns, when other conditions are equally favorable.

Then, again, soils are spoken of as warm or cold, having reference to their relative temperatures, especially during the early part of the season. And here the strongest factor in determining the temperature of a soil is the amount of water it can hold and bring to the surface for evaporation, those holding the most water and delivering it most rapidly at the surface being usually the coldest, and why this is so will be explained in another place.

Soils are sandy when not more than 40 to 65 per cent of their weight is made up of particles so small that from 1000 to 400,000 of them must be placed in line to span a linear inch, while the balance may be so large that only 20 to 100 of them are needed to stretch across the same distance. The heaviest clay soils, on the other hand, may have 80 to 95 per cent of their weights made up of the smallest sizes of particles named above, while only 5 to 20 per cent of grains of the larger diameters are found among them. Loamy soils are such as have their grains intermediate between those of the sandy and heavy clay types; while between this medium soil we have on the coarser-grained side, sandy loams and loamy sands, and on the finer-grained side, clayey loams and loamy clays, there being, of course, an insensible shading of one of these types of soil into another.

It has long been observed that these different types of soils in all parts of the world are regularly inhabited by kinds of plants peculiar to each, and by referring to the descriptions of soils on pages 82 and 83, it will be seen how sharply contrasted the types of vegetation are in that particular region. In the battle of the forests, in the race for sunshine and for moisture to meet the needs of thirsty leaves, some plants have suited their roots to the coarser and dryer soils, so that they feed themselves here more economically than others can; while other plants and other trees, in the long years of fitting and refitting, have come to thrive better than their competitors on the heavy clay soils. Nature, in her system of evolution, has developed a division of labor among plants just as she has among animals. She saw, long ago, that in order that sunshine might bring the largest returns out of soil, air, and water, there must be special adaptations to a limited set of conditions, and just as woodpeckers are a family of birds peculiarly fitted to gleaning food from the trunks of trees, so there are plants peculiarly adapted to the different types of soil.

Besides the soils just referred to there are the swamp, muck, peat or humus soils, which contain a very high per cent of decaying organic matter or humus, before referred to but these only occur in wet, undrained localities.

THE STORE OF PLANT FOOD.

Let us look now to the relation between the amount and kinds of mineral food materials stored in the soil and that removed by crops. Professor Wolff gives the composition of some fresh, or air-dry, agricultural products as follows: —

	MAIZE.		OATS.		WINTER WHEAT.		WINTER RYE.		BARLEY.		RED CLOVER.	
	Stalk.	Grain.	Stalk.	Grain.	Stalk.	Grain.	Stalk.	Grain.	Stalk.	Grain.	Stalk.	Grain.
Total ash . . .	47.2	12.3	44.0	26.4	42.6	17.7	40.7	17.3	43.9	21.8	56.5	36.9
Potash . . .	16.6	3.3	9.7	4.2	4.9	5.5	7.6	5.4	9.3	4.8	19.5	13.8
Soda	0.5	0.2	2.3	1.0	1.2	0.6	1.3	0.3	2.0	0.6	0.9	0.2
Magnesia . . .	2.6	1.8	1.8	1.8	1.1	2.2	1.3	1.9	1.1	1.8	6.9	4.5
Lime	5.0	0.3	3.6	1.0	2.6	0.6	3.1	0.5	3.3	0.5	19.2	2.3
Phosphoric acid	3.8	5.5	1.8	5.5	2.3	8.2	1.9	8.2	1.9	7.2	5.6	12.4
Sulfuric acid .	2.5	0.1	1.5	0.4	1.2	0.4	0.8	0.4	1.6	0.5	1.7	1.7
Silica	17.9	0.3	21.2	12.3	28.2	0.3	23.7	0.3	23.6	5.9	1.5	0.9
Sulfur	3.9	1.2	1.7	1.7	1.6	1.5	0.9	1.7	1.3	1.4	2.1	

From this table it appears that each 1000 pounds of air-dry product requires from the soil the number of pounds of each ingredient there indicated. Taking clover as an illustration, each ton of clover hay demands for its production 39 pounds of potash, 1.8 pounds of soda, 13.8 pounds of magnesia, 38.4 pounds of lime, 11.2 pounds of phosphoric acid, and 3.4 pounds of sulfuric acid; and two tons of hay per acre makes the annual demand double these amounts. What has the soil to offer?

If we take the mean dry weight of a surface foot of soil at 80 pounds and the mean soluble ingredients of soils at the percentages given by Hilgard on pages 84–87, regarding these as indicating the available plant food, then the amounts in the surface foot of an acre of land may be shown to be as follows: —

Potash	in surface foot, per acre	. .	3.76 tons.
Soda	" " "	. .	1.58 "
Magnesia	" " "	. .	3.92 "
Lime	" " "	. .	1.88 "
Phosphoric acid	" " "	. .	1.97 "
Sulfuric acid	" " "	. .	.91 "
Soluble silica	" " "	. .	73.40 "

It is not unfair to assume that every young man who starts on a farm of his own may be interested in its maintenance for 50 years on his own account and for 50 more years for the sake of his children. What then are the demands on the soil for 100 years likely to be? To give definiteness and simplicity rather than exactness to our problem, let us assume that a three-year rotation of corn, clover, and oats shall be adhered to throughout; that the clover shall all be cut as hay and with the straw and corn stalks, or their equivalents, constitute a revolving fund, never to go permanently off the farm; while the equivalent of the grain in some form shall be sold and no part returned to the land. Assume, further, that each and every year the yields shall be exactly two tons per acre of air-dry product, and that in the case of the corn and oats one-half of the product shall be grain and the other stalks or straw. Under these conditions how long will the several credits — potash, soda, magnesia, lime, phosphoric acid, sulfuric acid, and silica — last?

On the basis of these assumptions the amounts of the several ash ingredients required to be extracted from the soil by each of the three crops during the 33 years out of 99 which it will have fed upon the ground would be, —

FOR THE REVOLVING FUND.

	Potash.	Soda.	Magnesia.	Lime.	Phosphoric Acid.	Sulfur as Sulfuric Acid.	Silica.
	lbs.	lbs.	lbs.	lbs.	lbs.	lbs.	lbs.
Clover	2574.0	118.8	910.8	2534.0	739.2	536.5	198.0
Oat straw . . .	640.0	151.8	118.8	237.6	118.8	442.6	1399.0
Corn stalks . .	1095.5	33.0	179.7	330.0	250.8	953.0	1181.5
Total in 100 yrs.	4309.5	303.6	1209.3	3101.6	1108.8	1932.1	2778.5

FOR THE EQUIVALENT OF GRAIN SOLD.

Corn	217.8	13.2	118.8	19.8	363.0	249.2	19.8
Oats	277.2	66.0	118.8	66.0	363.1	370.0	812.0
Total in 100 yrs.	495.0	79.2	237.6	85.8	726.1	619.2	831.8

If we now divide the amounts of the several ash ingredients which chemical analysis points out as existing in the surface foot of the average soil of the humid regions, included in Hilgard's studies, by the amounts of these ingredients supposed to be sold and hence removed permanently from the land, the results will indicate the number of years required for the removal of the amount of the several ash ingredients standing to the credit of the surface foot of soil at the beginning of the 100 years. Making these divisions we get,

Potash	enough to last	1521	years.
Soda	" "	4050	"
Magnesia	" "	3300	"
Lime	" "	4387	"
Phosphoric acid	" "	542	"
Sulfuric acid	" "	292	"
Soluble silica	" "	17650	"

When it is observed that cultivated plants send their roots foraging through not less than the upper four feet of soil, and that the amounts of ash ingredients which have been considered are only those which chemical analysis tells us are to be found in the surface foot, it seems well-nigh impossible that there could ever be a deficiency in any one of the ash ingredients which plants find essential to their well-being.

Notwithstanding the apparently inexhaustible stores of potash, magnesia, lime, phosphoric acid, etc., in the soil, it must be remembered that we are nevertheless confronted with the indisputable results of practical field work which show, very often at least, that the addition of purely mineral fertilizers have been associated with larger yields per acre. We are confronted on every hand, too, with the fact that lands do run out and some varieties much more quickly than others; so that, when all has been said, the most important fact to bear in mind is that here is a problem lying at the very foundations of agriculture upon which a vast work has yet to be done.

In support of this view let me place in evidence some of the results obtained by Sir J. B. Lawes and his associates during their classic experiments bearing upon the conditions which determine the fertility of soils. By growing various crops year after year on the same land, to which no nitrogen-bearing manures were applied, they found that when purely mineral fertilizers were added, the crops were able to produce larger yields and to extract more nitrogen from the soil than when these fertilizers were omitted. Wheat, for example, grown continuously on the same land for 32 years without manures of any kind, was able to extract from the soil and build into its product 20.7 pounds of nitrogen per acre, per annum, but the same crop, on adjacent and similar lands, to which mineral fertilizers without nitrogen were added, was able to gather and store 22.1 pounds, an amount 6.76 per cent larger. Barley, during 24 consecutive years, drew from the soil 18.3 pounds of nitrogen per annum, where no mineral fertilizers were added, but 22.4 pounds per acre under the stimulus from them. Beans gathered from their land 31.3 pounds of nitrogen per acre where no fertilizer was added, as an average of 24 years, but 45.5 pounds where complex mineral food was given them. Ground bearing 6 crops of clover in 22 years, with 1 of wheat, 3 of barley, and 12 years fallow ground, gave, without fertilizers, 30.5 pounds of nitrogen per acre, but with the mineral fertilizers, 39.8 pounds. So, too, in a rotation of crops, 7 courses in 28 years, no fertilizers, gave 36.8 pounds, while with superphosphate of lime the yield was 45.2 pounds per acre. Then, again, in the mixed herbage of grass land, 20 years without fertilizers gave a mean yield of 33 pounds, but with a mixed mineral fertilizer, containing potash, the product of nitrogen was 55.6 pounds per acre, per annum.

Such results as these place beyond question the fact that, in spite of the large stores in the soil of all the ash ingredients used by plants, the addition of mineral salts more or less closely allied to those found there do give to the plant an added power over the natural resources of the soil in which they may be growing; and the marvelous part of the whole problem is that so small an amount of the fertilizer, when diluted by so much soil and water, can exert the appreciable effects observed.

CHAPTER III.

NITROGEN OF THE SOIL.

In speaking of the chemical composition of the soil, very little has thus far been said regarding nitrogen, the most important, perhaps, of all the ingredients; and for the reason that the loss of it from the soil is so rapid under faulty management, and the demands for it so urgent, as to claim for its consideration a separate chapter.

THE STORE OF SOIL NITROGEN.

We have seen that the total amount of the ash ingredients of plant food stored in the surface four feet of soil is very large, and so, too, is the total nitrogen, when considered in all its combinations. Storer, citing the observations of Krocker, to the effect that cultivated soils rarely contain less than .1 per cent of nitrogen, calculates the amount stored in the surface foot as seldom less than 3500 pounds per acre. This is an amount so large that if we take Warington's estimate of the nitrogen removed from the soil by wheat, when producing 30 bushel per acre, there is enough for more than 70 such crops, counting 33 pounds of nitrogen in the grain and 15 pounds in the straw. The mean amount of nitrogen in eleven arable and grass soils at Rothamsted is placed by Lawes and Gilbert at .149 per cent, and for eight other Great Britain soils at .166 per cent. Four Illinois prairie soils were shown by Voelcker to contain .308 per cent; while seven rich Russian soils have been shown by C. Schmidt to contain .341 per cent of nitrogen. If we bring all of these thirty analyses together into one mean, we shall have .219 per cent, an amount more than double that used by Storer as stated above.

In studying the fertility of the rich prairie soils of Manitoba, Lawes and Gilbert found that the surface foot, from four different localities, contained an average of .373 per cent of nitrogen, or enough, could it all be utilized in wheat production, for some 6500 bushels from each and every acre of land.

But the nitrogen of these and other soils is not all contained in the surface foot, and the distribution of it in the surface four feet of the soils in question, as they learned through chemical analysis, is given in the table below, which shows that the total amount is more than as large again as that given for the surface foot.

AMOUNT OF NITROGEN PER ACRE IN SURFACE FOUR FEET OF FOUR MANITOBA SOILS.

Depth.	Niverville. lbs.	Brandon. lbs.	Selkirk. lbs.	Winnipeg. lbs.
1st foot	7308	5236	17304	11984
2d foot	5408	3488	8448	10464
3d foot	2484	2592	2736	5688
4th foot	1520	870	1487	4045
Total	16720	12186	29975	32181

Nor can it be urged with these soils that the nitrogen is in a form which cannot, under favorable conditions of tillage, be converted into nitric acid and nitrates, the form in which nitrogen is largely used by cultivated plants; for Sir J. B. Lawes caused samples of these soils to be placed under favorable conditions for nitrification, where they were held, with some interruptions, during something more than 300 days and during this time it is estimated that not less than 799.8 pounds per acre, as an average for the four soils, were converted into soluble nitrates available for crop production, as shown in the table which follows: —

AMOUNT OF NITROGEN PER ACRE CONVERTED INTO SOLUBLE FORM.

Depth.	Mean Per Cent of Total Nitrogen.	Niverville. lbs.	Brandon. lbs.	Selkirk. lbs.	Winnipeg. lbs.
1st foot	4.295	234.1	250.4	655.1	650.0
2d foot	2.733	72.9	57.2	377.5	372.5
3d foot	2.108	12.6	53.7	27.2	276.0
4th foot	1.443	15.0	10.7	11.5	112.8
Total		334.6	372.0	1071.3	1411.3

While the conditions under which these samples were placed for nitrification were much more favorable for the rapid carrying forward of the process than could be produced in the natural soil, it should not be inferred that the amount so changed represents the total available nitrogen in these soils; for there appears to be no reason why time and favorable conditions may not transform into soluble nitrates much the larger portion of that which the analyses have shown them to contain.

The presence of these large amounts of nitrogen and of ash ingredients which soils have been shown to possess are very fundamental facts, which must be kept ever in mind in our efforts to restore the vigor of exhausted soils and to maintain a high degree of productiveness in them when so brought back. They are important because they show very clearly that run-out lands, and lands tired of this crop or of that, have been rendered so by some change different from the consumption of the chemical ingredients contained in soils out of which plant food is elaborated. They are important because they bring clearly into view the fact that, as yet, our practices regarding the rotation of crops, regarding the fertilizers or manures for this soil or that crop, and regarding many of our efforts to improve our lands by this or that method of tillage, are all of them far too much like those of a man feeling his way in the dark. They are important because they show that the agriculture which shall be able to feed the crowded cities, so rapidly building, as cheaply as they must be fed, must bring to its aid the mightiest efforts of the strongest minds, and must place the management of its lands under trained men whose education is unexcelled in its kind by that of any other trade or profession.

IMPORTANCE OF NITROGEN IN PLANT LIFE.

Nitrogen, sulfur, and phosphorus are the three elements, derived from the soil, which are largely peculiar to those highly complex compounds so immediately associated with the life processes of plants. It is true that oxygen, carbon, and hydrogen, combined to form cellulose, out of which the framework of vegetation is built, and to form the starches, sugars, and gums, make up by far the largest part of the dry weight of vegetable matter; but these appear to be only products which have resulted from the transformations which take place within the protoplasmic substance of which nitrogen, sulfur, and, at times, phosphorus appear to be the controlling or life-leading elements. While carbon, oxygen, and hydrogen make up the larger part by weight of these responsive substances, yet by themselves they build into molecules too strong and too stable to be again thrown down by sunshine alone. Like the stick, which constitutes both the bulk and the chief source of heat in the common match, but a useless thing with which to kindle a fire unless tipped with the easily upset molecules of phosphorus, sulfur and nitre, which by their fall produce the necessary heat to fire the bit of wood, so the starches, the gums, and the fats stored in the seeds, in the fleshy roots and thickened stems and leaves of plants, could not serve as food and become sources of power, either in plants or in animals, were there not associated with them enough of those proteid or albuminoid molecules containing nitrogen, sulfur, and perhaps also phosphorus which, after absorbing a little too much water, are easily overturned by the warmth of the soil, and thus in their fall set on foot the train of changes it is their mission to start and to maintain.

It is very suggestive and not a little strange, in this connection, that one of the most powerful explosives we have is now made by slipping out of cellulose three of its hydrogen atoms and placing in their stead three groups of two atoms taken from nitric acid, this result being accomplished by simply steeping cotton wool in strong nitric acid, made more active by mixing with it twice its volume of concentrated sulfuric acid. When this is done and the wool has been washed and dried, it is found that, without having lost even its fibrous structure, the taking-out and slipping-in process has resulted in so unsteady a structure that a trifling disturbance causes it to fall apart with explosive violence. So, too, when glycerine is treated in a similar manner, a slipping-out and slipping-in change takes place; three bricks, so to speak,—three atoms of hydrogen,—are replaced by three others derived from the nitric acid, and instead of the oily, sweetish liquid, glycerine, we have the fearful nitroglycerine which, when mixed with earth, constitutes dynamite. Even in our old explosive, gunpowder, we employ the same nitric acid in combination with potash as saltpetre, which, with the addition of a trifling amount of heat, is made to give over its oxygen to the carbon and sulfur with which it is mixed, and, by uniting with them, almost instantaneously develops the power sought.

Now we are not trying to point out here just how the energy manifested in the tissues of plants and in the bodies of animals is derived: we are rather trying to point out how indispensable to the life of plants nitrogen is, that we may the more fully feel the urgency of holding its amount and condition in the soils we till up to the standard demanded by vigorous plant growth. All are agreed that, varied as the constructive and destructive processes of plant life are, they take place within or are directly or indirectly set up by, the molecular shiftings, adjustments, and readjustments within the protoplasmic substance of cell life, of which nitrogen is an indispensable ingredient.

RELATION OF SOIL NITROGEN TO CROP DEMANDS.

Just how is the amount of nitrogen removed by a crop from the soil related quantitively to the ash ingredients, and how are the amounts of these in the soil related to the several amounts in the crop? The crop of wheat which yields 30 bushels of grain per acre demands, as indicated by chemical analysis, 48 pounds of nitrogen, 21.1 pounds of phosphoric acid, 28.8 pounds of potash, 9.2 pounds of lime, 7.1 of magnesia, 7.8 of sulfur, and 96.9 pounds of silica. A crop of oats yielding 60 bushels per acre would demand: of nitrogen, 73.3 pounds; phosphoric acid, 25.7 pounds; potash, 61.5 pounds; lime, 15 pounds; magnesia, 11.6 pounds; sulfur, 10.7 pounds; and of silica, 113.7 pounds. In the case of red clover, yielding 2 tons of hay per acre, the demands are: for nitrogen, 102 pounds; phosphoric acid, 24.9 pounds; potash, 83.4 pounds; lime, 90.1 pounds; magnesia, 28.2 pounds; sulfur, 9.4 pounds; and silica, 7 pounds. These are the amounts of plant food, of the kinds named, demanded from each acre of ground in the cases cited.

Taking the average per cent of nitrogen in the surface foot of soil at .15 per cent, the weight of the soil at 80 pounds per cubic foot, and the ash ingredients as indicated on page 101, it will be seen, after dividing the amounts of the several ingredients in the soil by the respective amounts demanded in the three cases, that, could the materials be all used, there would be of

		Wheat. Crops.	Oats. Crops.	Clover. Crops.
Nitrogen,	enough for	109	72	52
Phosphoric acid,	" "	187	153	158
Potash,	" "	261	122	90
Lime,	" "	409	251	42
Magnesia,	" "	1104	676	278
Sulfur,	" "	76	56	63
Soluble silica,	" "	1515	1291	20970

In these figures it is assumed that nothing is given

back to the soil by the crops, which, however, is not true regarding nitrogen, especially with the clover. It will be seen from the figures that the supply of nitrogen is relatively less than that of any other ingredient except sulfur.

If it were true that all of these plant foods are both equally available and necessary, and the possibilities of loss the same, then the tendency would be for the nitrogen and sulfur to give out earliest. Now, in the case of nitrogen, certainly no plant food from the soil is more important and probably no one is subject to as great loss under bad management as this; and hence it follows that much attention should be paid to this soil ingredient.

FORMS OF NITROGEN IN THE SOIL.

Nitrogen occurs in the soil under several different conditions, and is derived from several sources. As a part of the soil air, it exists as free nitrogen where it is essential to the life of certain microscopic forms which seize upon it and render it available to higher plants. Temporarily and in transition stages, nitrogen occurs in the soil as ammonia and as nitrous acid, but these forms pass rapidly, apparently, into nitric acid, from which most of the higher plants derive their chief supply. The amount, therefore, of ammonia or of nitrous acid existing in a soil at any time is always small. In the form of nitric acid, however, the soil contains considerable amounts of nitrogen, as nitrates of lime or other bases, but in this form Warington states that the amount seldom reaches 5 per cent of the total nitrogen content.

By far the larger part of the nitrogen of soils is stored in the form of humus, to which reference has already been made; and this, through the processes of fermentation, is gradually made available to plants in the form of nitric acid. The roots of field crops, too, contribute no small amount of material to the store of organic matter in the soil, a part of which is converted into humus; but the plants whose roots leave in the soil the largest amounts of nitrogen are the leguminous species, like the clovers, beans, peas, and lupines.

DISTRIBUTION OF NITROGEN IN THE SOIL.

Regarding the distribution of nitrogen in the soil, some facts have already been given on page 108. Referring to the table, it will be seen that the amount decreases downward until in the fourth foot the average is less than one-fifth of that found in the first foot. This decrease in the amount of nitrogen is evidently due in part to the variation in the distribution of roots in the soil, and perhaps largely to this cause; but there are other factors which tend sometimes to cause a destruction of the nitric acid, setting free the nitrogen in the gaseous form. Observation has shown that when a soil rich in nitrates is saturated with water so as to exclude free oxygen, a deoxidation may take place which sets free nitrogen gas; nor is this action limited to the nitrates, for organic matter is also broken down under these conditions, with a setting free of both carbon dioxide and nitrogen gas.

The distribution of nitrogen in Rothamsted soils is given by Warington to a depth of 9 feet, and the results are quoted as follows: —

NITROGEN IN SOILS AT VARIOUS DEPTHS.

	Arable Soils. lbs. per acre.	Old Pasture Soils. lbs. per acre.
First 9 inches contained	3015	5351
Second 9 " "	1629	2313
Third 9 " "	1461	1580
Fourth 9 " "	1228	1412
Surface 3 feet contained	7333	10656
Fifth 9 inches contained	1090	1301
Sixth 9 " "	1131	1186
Seventh 9 " "	1049	
Eighth 9 " "	1095	
Second 3 feet contained	4365	
Ninth 9 inches contained	1173	
Tenth 9 " "	1076	
Eleventh 9 inches "	1112	
Twelfth 9 " "	1198	
Third 3 feet contained	4559	
Surface 9 feet contained	16257	

It will be seen from this table that the amount of nitrogen in an old arable soil may be really very large, in this case enough, could it all be used, for 338 crops of wheat of 30 bushels per acre, or more than 10,000 bushels. It will be seen also that, after the first three feet are past, the total amount of nitrogen in the succeeding feet remains nearly constant, or that it tends very slightly to increase. The figures suggest that at a certain depth below the surface of arable fields, the amount of nitrogen tends to remain approximately constant from year to year, while the rise and fall of this ingredient is a phenomenon largely peculiar to the surface three or four feet.

Referring now to the distribution in the soil of the nitrates simply, it must be said that the amounts of these found at different depths varies with the season and to some extent also with the crop. At the time crops are growing, the roots remove the nitrates, where these roots are found, so thoroughly that the measured amount at any one place is small, the plants tending to pick it up as rapidly as it may be formed. At the end of a season, when percolation has been taking place for some time, the nitrates are largely removed in the drainage waters, or they are carried far below the surface. So too, if a considerable quantity of nitrates have newly formed near the surface and there comes a heavy rain which produces percolation from the first foot into the second foot of soil, the effect of this percolation is to shove along in its front layer of moving water nearly all of the readily soluble salts the soil may have contained, thus changing their position from the first to the second foot in perhaps one or two days. Such action as this tends to concentrate the nitrates in a thin zone and at times possibly wholly below the territory occupied by the roots. Such concentrations, however, cannot be permanent; for both diffusion and capillarity tend to bring about a redistribution again.

SOURCES OF SOIL NITROGEN.

When we ask, From whence comes the nitrogen of the soil? we raise one of the most important questions of practical agriculture, and one which is to-day taxing, to the utmost, the skill of many very able investigators. When the earlier investigators upon this subject found their results, one after another and almost without exception, pointing to one conclusion, namely, that the plants upon which they experimented either did not increase the total nitrogen given them or where an increase appeared to be associated with their growth this was so small, or of such a character, that it seemed more than likely it might have been derived in some way outside of the direct action of the plant under experiment, it came to be almost accepted that the vegetable world depended for its nitrogen upon the decay of organic substances supplemented in small part by the nitric acid and ammonia which are brought down by the rains and the snows or condensed in the dews. It was indeed early observed that clover could be grown upon land relatively poor in nitrogen, and at the same time remove in the crop produced large amounts of that ingredient, while a crop of wheat following the clover was able to procure more nitrogen from the same field, following the clover, than it was able to do before the clover had occupied the land and taken from it its large amounts of nitrogen. We now know that the growth of clover upon land may be associated with the accession from the atmosphere of large amounts of nitrogen, but the studies of the earlier investigators prevented them from believing such an explanation possible, and various hypotheses were advanced to explain the advantages derived from systems of rotation in which clover was one of the series.

The fact that nitrates were early known to be drained away from the land in large quantities and that nitrogen gas is liberated during the processes of decay, when coupled with the fact that life has existed upon the land during untold geologic ages, should have made it appear inevitable that there must be some way for drawing from the free nitrogen of the air a supply sufficient to make good the losses referred to, and especially when it was known that the rocks themselves, from which soil is derived, do not contain the element in appreciable quantities. And then whenever the question was asked, From whence came the nitrogen which entered into the very earliest forms of life and which made the very earliest soils fertile? it should have seemed almost axiomatic that the nitrogen of the air must be used over and over again, as we now know, but only through the studies of the last decade, that it in reality is.

The amounts of nitrogen brought to the soil in the form of nitric acid and ammonia, with the moisture precipitated as rain, snow, and dew, are known to be relatively small, rarely more than the equivalent of 5 pounds per acre per annum in the open country removed from the influence of the burning fuel and products of sewage decomposition associated with the life of large cities. A knowledge of the actual amounts of nitrogen so derived has been gained by keeping records of the total precipitation and by determining the amounts of nitrogen such waters did actually contain. Some of the results which were obtained through careful studies at Rothamsted are given in the following table:—

NITROGEN AS AMMONIA AND NITRIC ACID, IN POUNDS PER ACRE PER ANNUM, IN RAIN.

	ROTHAMSTED. 8 years. lbs.	LINCOLN, NEW ZEALAND. 3 years. lbs.	BARBADOES. 3 years. lbs.
Nitrogen as ammonia	2.53	0.74	0.93
Nitrogen as nitric acid	0.84	1.00	2.84
	3.37	1.74	3.77

These amounts, it will be seen, are only sufficient to contribute the necessary nitrogen for about two bushels of wheat per acre where 48 pounds is counted sufficient for 30 bushels of wheat.

Observations similar to the above have been made in various parts of Europe, and the mean of 22 determinations, each extending over a whole year, at nine different stations, give results varying from 1.86 pounds of nitrogen per acre to 20.91 pounds, with a mean value of 10.23 pounds per acre where the average amount of rainfall was 27.03 inches. From these figures it will be readily seen that even were any country to receive annually supplies of nitrogen equal to the largest amount here stated, the quantity would be insufficient for continuous large crops, even could the whole of it be used for that purpose.

Before leaving this subject, — the contributions of rain to the soil, — let us quote a table from Dr. Angus Smith in "Air and Rain," in which the average composition of rain water from various parts of England and Scotland are given.

AVERAGE COMPOSITION OF SAMPLES OF RAIN IN PARTS PER MILLION.

	Nitrogen as		Chlorine.	Sulfuric Acid.
	Ammonia.	Nitric Acid.		
England, country places, inland	.88	.19	3.88	5.52
Towns	4.25	.22	8.46	34.27
Scotland, country places, seacoast	.61	.11	12.24	5.64
Inland	.44	.08	3.28	2.06
Towns	3.15	.30	5.70	16.50
Glasgow	7.49	.63	8.72	70.19

From this table it will be seen the rains bring down larger amounts, both of chlorine and of sulfuric acid than of nitrogen in both of its combinations. It will be observed that in and about towns and cities all of the ingredients are more abundant than they are in the open country, and that the chlorine, which occurs in the air in the form of common salt, is more abundant near the sea than it is farther inland, as should be expected because it is largely derived from sea-water. The salt does not evaporate, but in times of high winds, when the water on the crests of the waves and breakers is whipped into fine spray, the water so taken into the air is evaporated, leaving the salt floating in the air as fine particles of dust about which the raindrops form and so bring them to the ground. Since England and Scotland together form a relatively small body of land entirely surrounded by salt water, it is probable that both the amounts of salt and of sulfuric acid, even from the country places inland, are larger than are likely to be observed in the interior of continental areas. The large differences between the cities and the country, shown in the table, are due chiefly to products of combustion, issuing from the many stoves, furnaces, and engines which are concentrated within their borders.

If we take the average of the sulfuric acid from the inland country places in England and Scotland and calculate the amount of sulfur per acre with different depths of rainfall, and then set these amounts alongside of the amounts of sulfur which chemical analysis has shown to occur in crops of wheat, oats, and clover, as given on page 101, we shall have the results which follow:—

With rain containing 3.79 parts of sulfuric acid per million, —

SULFUR.

12 inches of rain would yield	.89	pounds per acre.
24 " " " " "	1.78	" " "
36 " " " " "	2.66	" " "
30 bushels of wheat demand	7.8	" " "
60 bushels of oats demand	10.7	" " "
2 tons of clover hay demand	9.4	" " "

It appears therefore, from these results, that even with 36 inches of rain per annum, less than one-third of the necessary amount of sulfur needed by large yields of crops per acre can be supplied by the sulfuric acid of rains.

It is held by some that gaseous ammonium carbonate passing in the air over the foliage of vegetation may be absorbed by the leaves, and thus contribute to the supply of nitrogen for the purposes of plant growth; but upon this point the evidence appears to be insufficient even to place beyond question the fact of such a source of nitrogen, much less to allow of a quantitative expression of its amount.

Schlösing, however, in his experiments on moist soils freely exposed to the air, believes he has shown that under certain conditions soils may acquire nitrogen, through such an exposure, at the rate of 38 pounds per acre per annum, and the additions to his experimental soils appear to have been largely in the form of ammonia. It will be readily understood that if soils do have the power of removing from the passing air the compounds of nitrogen it contains, the amounts of nitrogen added to the soil in the combined form might easily be more than that which is brought down by the rains, because the steady ongoing of the winds across the surface of the land must necessarily bring a very large amount of air in contact with a given field of soil during the course of any season. Still, to obtain 38 pounds of nitrogen from the air in the form of ammonia, means the complete removal of this gas from a very large volume of air.

The daily analysis of air on Montsouris from 1876 to 1881 gave the mean quantity of ammonia in the air at 2.2 milligrams for each 100 cubic metres, and for a soil to acquire 38 pounds of nitrogen from the ammonia, it must completely extract from the air that which is contained in 951.4 millions of cubic metres. This means that if, on an acre, the air one metre deep, which is a little more than 3 feet, were to give up all of its ammonia to the soil, that air would have to be changed more than 2.5 million times in order to bring the necessary amount of ammonia to the acre to yield 38 pounds of nitrogen. These and other considerations make it appear quite certain that soils, under average field conditions, cannot receive annually the above amount of nitrogen in this form from the air, at least outside of tropical regions.

It is well established that electrical discharges in the atmosphere cause some of the oxygen of the air to unite with nitrogen, giving rise to the nitrous and nitric acids which the air is known to contain and which have been referred to as being brought down by the rain. But it is probable that a part of these combinations with oxygen are derived from ammonia through the action of ozone, a modified and highly active form of oxygen. This action is believed by Warington to be supplemented by the peroxide of hydrogen, and hence that the nitrates brought down in the rain do not represent so much nitrogen not previously combined, but rather that a portion of these may have resulted from the oxidation of ammonia.

FREE-NITROGEN-FIXING GERMS.

We come now to consider one of the most important discoveries of modern times; a discovery which places in the hands of every farmer a means, completely under his control, whereby he can at any time cause to be drawn directly from the atmosphere the free nitrogen of the air and have it fixed in the soil of any field he may wish to enrich.

The history of this important discovery cannot be given here, but in 1888 Hellriegel published the results of his experiments, which satisfactorily proved that we have in nature great numbers of minute microscopic organisms living in the soil, which have the habit of locating themselves upon the roots of certain kinds of plants and especially upon the clovers, beans, peas, lupines, and other leguminous species. These organisms, once seated upon the roots of a plant congenial to them, cause the formation of little enlargements or tubercles, in which they live, drawing nourishment from the plant upon which they have established their home, but in return giving to it compounds of nitrogen which these organisms are able to produce from the free nitrogen of the soil air. Here we have two neighbors changing work; each performing that sort of labor which, through long years of specialization, it has become best fitted to do, and then swapping the products of their labor to the mutual advantage of both. The clover, with its leaves outspread in the free air and fitted to utilize the bright sunshine as a source of power, breaks down the molecules of carbon dioxide, forging them into such compounds as it needs; but in its efforts to utilize to the best advantage all of its surroundings, it has given over the first rough forging of nitrogen into food to the lowly organisms upon its roots, and now sends down to them so much of such compounds best brought together under the hammer of direct ether waves as they need in doing their part of the work. Like the coal brought up from the mine, like the fuel in the engine, the products made in the green leaf and sent down to the lightless cellars become the source of power there used by the one-celled plants in bringing into combination with oxygen the free nitrogen of the air, so indispensable to the life of higher plants. What a community of life! What a complete utilization of all forces and all materials, both above and below ground, we have here!

If the roots of clover plants are examined, there may easily be seen upon their surfaces many little lumps, knobs, or tubercles. Similar bodies may also be observed upon the roots of peas, beans, lupines, and other leguminous plants, and these are the places where colonies of the bacteria here under consideration have located themselves. These micro-organisms are found scattered through the soil, and some believe that they have the power, at least certain species, of withdrawing free nitrogen from the air without the aid of another plant on which to live, but the matter still awaits positive demonstration.

Some of Hellriegel's studies make it appear that different species of leguminous plants are inhabited by varieties of these minute organisms peculiar to themselves. He found, for example, in preparing his sandy soils free from nitrogen, in which he proposed to grow lupines, that, in order to have these plants succeed, it was necessary to wet the sands with water leached from lands where these plants had habitually grown. The application of water from rich loams which had produced peas, beans, and vetches appeared to do them no good; the inference being that these soils did not contain the bacteria which live upon their roots, while the natural lupine soils are rich in them. Experiments have been made in introducing the bacteria from the tubercles from pea roots directly into the roots of lupines, but without

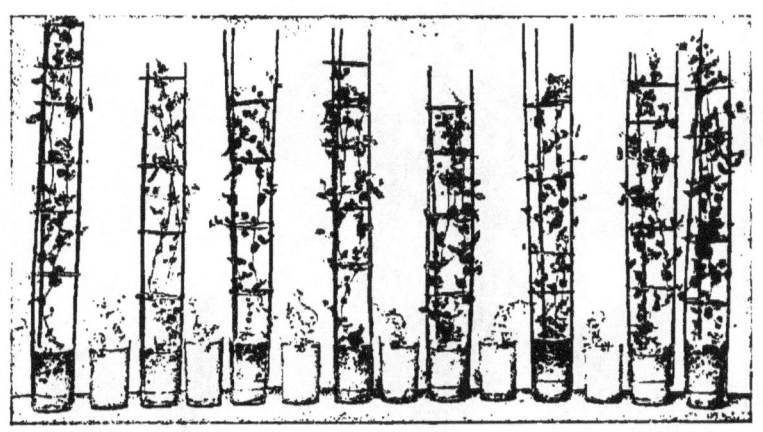

FIG. 18. — Showing the influence of free-nitrogen-fixing germs on the growth of peas. The large plants all grew in sand containing the nitrogen-fixing bacteria, while the small plants grew in soils identically the same except that all bacteria were excluded from them. After Hellriegel.

proving of any help to them. A full study of these problems may lead to an explanation of why it is sometimes difficult at first to get clover to do well on certain soils, and also what is the cause of "clover-sick" lands.

To show how great is the service of these free-nitrogen-fixing germs, there is given below the results of eight trials made by Hellriegel with lupines, in which all of the plants were treated alike except that four of them were, at the start, given loam water in order to add the necessary bacteria, while the other four pots received none. Taking the amount of nitrogen in the plants forced to grow unaided by the nitrogen-fixing germs as 1, the amounts stand, —

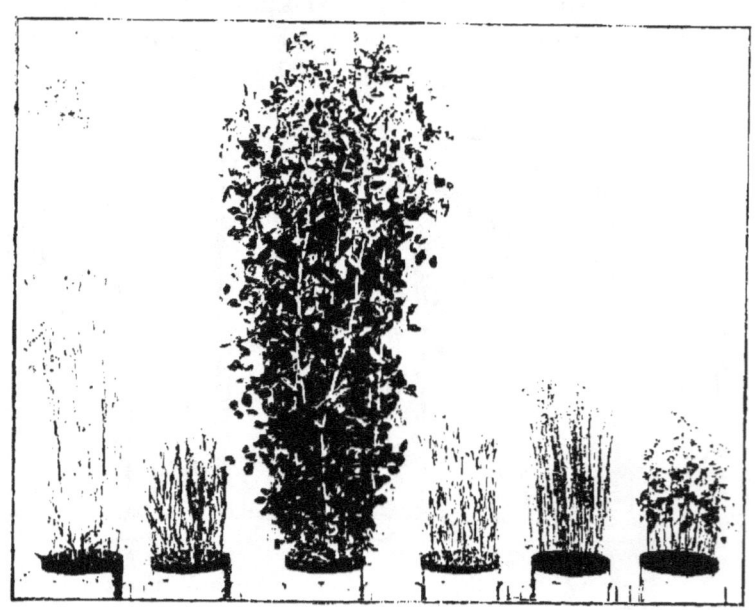

Fig. 19. — Showing the growth of rye, oats, peas, wheat, flax, and buckwheat in soils fertile in all elements of plant food except nitrogen, and illustrating the power of the pea, through its root tubercles, to procure nitrogen from the air. After P. Wagner.

	NITROGEN.		NITROGEN.
No. 1, without germs	1	No. 2, with germs	76.07
No. 3, " "	1	No. 4, " "	82.36
No. 5, " "	1	No. 6, " "	91.15
No. 7, " "	1	No. 8, " "	100.54

If, however, we compare the weights of dry matter produced, these stand, —

	Dry Matter.			Dry Matter.
No. 1, without germs	1	No. 2, with germs	48.66
No. 3, " "	1	No. 4, " "	57.01
No. 5, " "	1	No. 6, " "	48.30
No. 7, " "	1	No. 8, " "	41.58

These two comparisons bring into strong relief the great help the lupines received from the nitrogen-fixing germs which grow upon their roots; for they show that, while there is a strong difference between the amounts of dry matter produced under the two conditions, there is a much larger difference between the amounts of nitrogen in the crops.

The same facts are rendered still more emphatic in Fig. 18, which shows the results of growing peas in pots of soil in which all germs had first been killed, and then introducing them again in one set but not in the other. In Fig. 19, too, will be seen how differently the plant thrives which can get the nitrogen it needs from the air through the aid of these germs than do those not able to profit by their help, while Fig. 20 shows the importance of nitric nitrogen to other plants.

SYMBIOSIS.

This living together of plants for their mutual benefit or symbiosis, as it has been technically called, is not a rare occurrence. Indeed the whole family of lichens, already referred to, are now regarded as examples of this phenomenon, the forms being in reality fungus, puff-ball or toadstool-like plants densely inhabited by green or chlorophyll-bearing algæ. These two types of life live together to the advantage of both. And what is still more strange and interesting in this connection, is the observation that members of the animal kingdom, like the Radiolaria and Actiniæ, owe their colors in part to the many minute, one-celled algæ living within their bodies, and there consuming the carbon dioxide given off by the animal and in their turn producing starch, under the action of sunlight, which serves as food for their host.

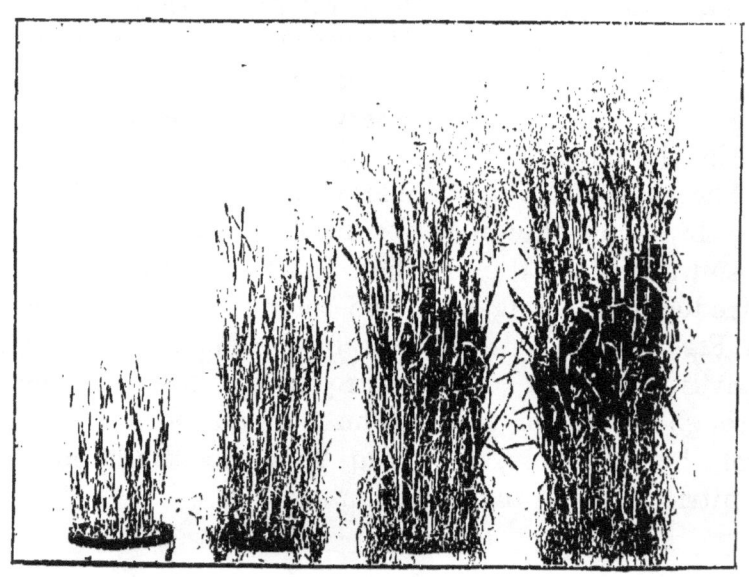

Fig. 20.—Showing oats growing under conditions identical with those of Fig. 19, except that the several pots received Chile saltpetre, 1, 2, and 3 grams respectively, thus enforcing the immense importance to such plants of nitric nitrogen. After P. Wagner.

Then, again, as a matter of immediate and practical importance to agriculture, it appears from the studies of Frank, and afterwards Schlösing, Jr., and Laurent in 1891, followed by Kosswitsch in 1894, that there are in the soil certain nitrifying bacteria which, living in symbiotic relation with soil algæ, are able to and do, under favorable conditions of light and moisture, fix considerable amounts of free nitrogen. Further than this, there is already at hand some evidence in favor of the view that even upon the soilless rock, which has become the abode of lichens, the process of fixing free nitrogen is there carried forward, if not through the double form of life which the lichen really is, then through the aid of a process similar to the one just mentioned.

After what has been said regarding these sources of nitrogen in the soil, it may be asked, How then can there ever occur a deficiency of it? The answer of course is that Nature, in her universal method of over and over again, has her counter processes of denitrification, of returning to the air from which it came the nitrogen, just as is the case with carbon dioxide; and man, when he sets his will and his ignorance against natural methods, has often given these latter agencies the ascendency. This brings us to consider how the nitrogen, once fixed in the soil, is used by plants, and how it may be lost.

NITRIFICATION AND DENITRIFICATION.

It has not yet been satisfactorily determined in what form the nitrogen fixed by the bacteria living upon the roots of plants, is used by those species, but the general belief, regarding plants whose roots are not inhabited by nitrogen-fixing germs, is that they are almost wholly, if not entirely, dependent for their nitrogen supplies upon the nitric acid, and possibly also to some extent upon ammonia in the soil. We need, therefore, to learn how these substances are brought to the soil, and how they may be lost.

We have already seen that some nitric acid and some ammonia come to the soil from the atmosphere directly, but in quantities too small to meet the demands for them. By far the larger part of the nitric acid found in soils is the final product of life processes there carried on when conditions are favorable for them to go forward.

All are familiar with the odor of ammonia as it rises from the heap of fermenting manure. This ammonia is produced from the compounds of nitrogen in plant tissues and in the excretions of animals through the action of certain kinds of bacteria. The same process, too, takes place in the soil where organic matter decays, and through the same agencies; but in the soil no sooner is the ammonia produced than it is seized upon by another organism, the nitrous ferment, and converted into nitrous acid. But the process does not end here; for no sooner is the nitrous acid formed than it is seized upon by still another distinct kind of micro-organism and converted into nitric acid, when it becomes available to higher plants.

It appears, therefore, from these facts, that we have in nature, so far as the nitrogen is concerned, a short-circuit rotation, this element of plant food not starting from its free, uncombined state in the air and passing through the compounds of living tissues and then back to gaseous nitrogen again in each and every round. On the contrary, there is a very large amount of nitrogen, in the form of nitric acid, being removed from the soil and transformed into the higher compounds of plant life, and then, after it has served its purposes in this form and the plants which have used it die, it passes back through ammonia and nitrous acid, reaching the nitric acid form again without having become nitrogen gas. Then too, when animals feed upon the higher plants, the circle in which nitrogen makes its round is widened, and it is rendered even still broader in those cases where flesh-eating animals place themselves in the circuit.

There are, however, processes going on in nature whereby the nitrates may be, for the time, entirely lost to higher plants and animals, these compounds being sometimes borne away in the drainage waters to the sea, there to serve the purposes of marine life. Then, again, there live within our soils microscopic forms of life which, when a rich soil is kept over wet and not well ventilated, are able to live, not only upon the nitrates, converting them into free nitrogen gas, thus impoverishing the soil, but also to feed upon the organic compounds as well, here, too, restoring free nitrogen to the air.

Dr. Angus Smith showed in 1867 that nitrates in sewage waters are decomposed with the liberation of free atmospheric nitrogen. Schlösing has also shown that a moist humic soil very quickly lost all traces of nitrates after being kept in an atmosphere devoid of free oxygen. Warington has shown that even sodium nitrate is decomposed in a water-logged soil, and a large part of the nitrogen set free as nitrogen gas. So great does the demand for oxygen appear to be in these water-soaked soils, that certain micro-organisms are even able, according to the observations of Müntz, to withdraw it from chlorates, bromates, and iodates, leaving in their stead chlorides, bromides, and iodides. So too it has been shown that when suitable forms of organic matter are brought into the presence of nitrates, even though some oxygen be there, the life demands for it may be so urgent that the nitrates may be broken down and nitrogen gas set free.

There is still another condition under which the processes of denitrification take place. It is the rapid, large, and almost complete loss of nitrogen from human excrement, where dry earth is used in the dry-earth closets. Storer quotes from various authorities a considerable amount of evidence relating to this phase of the subject. One of the most striking cases cited, is that where Colonel Waring kept two tons of dry earth for a number of years, having it used over and over again in order to see how long it might be employed without losing its efficiency. It is stated that the closets were filled with this dry earth about six times each year, and that when the vaults were emptied the earth was thrown into a heap in a well-ventilated cellar to dry. After this material had been used over not less than ten times, samples were sent to Professor Atwater for analysis. His examination showed that these ten-times-used soils contained no more than 11 pounds of nitrogen in the 4000 pounds of soil, and yet it is estimated that not less than 230 pounds of nitrogen had been added and that not less than 3 pounds existed in the soil when it was taken for use. It thus appears that all but 8 pounds of the 230 pounds of nitrogen added to this dry soil had been, during the interval, converted into gas. Nor was this all. So complete had been the destructive processes that nearly all the carbonaceous materials, including the paper used, had entirely disappeared.

Nature, then, so far as nitrogen compounds are concerned, just as with other matters, has her constructive and destructive processes pitted one against the other in such a manner that they tend all the while to balance. Go where you will, in almost any natural meadow or natural wood, and you will find growing there some plant of the leguminous species upon whose roots the free-nitrogen-fixing bacteria have their home, and where they are performing that service so indispensable to the higher types of life. Living with these nitrogen gatherers, there are, of course, other species which feed upon the nitrogen compounds produced by the plants referred to, while still other forms return to the atmosphere the nitrogen in its free and uncombined form. But man, not understanding how the nitrogen supply of the soil must be maintained, has shut out completely for many years in succession the symbiotic life referred to, and has attempted to grow crop after crop of wheat and other plants able only to consume the nitrogen of the soil without adding appreciable amounts to it, and with the inevitable result that the available nitrogen supply has fallen below the level needful to large yields.

CHAPTER IV.

CAPILLARITY, SOLUTION, DIFFUSION, AND OSMOSIS.

BEFORE taking up the physical problems of the soil, it will be very helpful if we can first get a clear conception of the processes of capillarity, solution, diffusion, and osmosis.

This is important, because they are the modes by which all the food of plants is conveyed to them, whether this food comes from the air or from the soil.

Whenever a lamp is lighted, there is at once started a current of oil, rising from the bowl through the many interspaces between the threads of the loosely woven wick, and this flow continues so long as the oil is removed from the upper end by burning. So, too, when a wet soil is exposed to a drying wind, the removal of water from its surface results in setting up a flow from deep in the ground, upward toward and to the drying surface, to make good these losses. Then, again, when the root of a plant threads its way among the soil grains and withdraws from their surfaces a portion of the water with which they are charged, there is at once set up, to make good this loss, currents of water traveling from various directions in the soil toward the absorbing root. By what mechanism are these movements maintained, and what is the source of energy by which the work is performed?

When the water rises from a well through the action of the pump, we know that the energy generated in the muscles of some man's arm has been transferred, through the pump handle and the piston, to the rising column of water; or if not the energy of human muscles, then that from the moving wind, acting through the windmill, or some other equally evident source of power. Matter is never moved, work is never performed except at the expense of energy in one form or another; so in the capillary movements of water through the soil, and in the rise of oil through the wick of a lamp, work is being done, and energy from some source must be expended.

SURFACE TENSION AND CAPILLARITY.

The first careful study of the rise of water in capillary tubes was made by Hauksbee nearly 200 years ago, but history shows that the phenomena were known to Leonardo da Vinci, the famous artist, who lived between 1452 and 1519. But notwithstanding the large amount of very careful study which these phenomena have received even during recent years, we are yet in the dark as to just how the energy which forces the capillary fluids to move is transformed into current motion; but all are agreed that it is in some way brought about through the surface tension of liquids.

During the blowing of soap bubbles, every one has observed that the bubble left hanging from the bowl of the pipe contracts, grows rapidly smaller, and forces a strong current of air out through the stem, showing that the thin film of water is acting like a stretched rubber ball. Now, if a large drop of water or a marble could be placed in the centre of the bubble, it is evident that the thin film would close down upon it, compressing it with so much force as it is able to exert, and in the case of the drop of water, the film would tend to keep it spherical in form. Many different phenomena show that the free surfaces of all liquids act upon the mass within or beneath them very much as if they were elastic and contracting membranes, and the spherical form of the raindrop, of the melted lead as it falls from the shot tower, of the dewdrop on the cabbage leaf, and the spherule of water as it rolls along the dusty floor, or glides from side to side on the surface of a hot stove, each and all owe their shape to this surface tension. It is this surface tension which allows a polished needle, although seven times heavier than water, to float easily upon its surface, and it is advantage of this, too, which the water spider takes as it glides easily along the water surface of ponds and slow streams.

This surface tension is a condition which results from the fact that all molecules in the surface of a liquid are being pulled more strongly into the liquid, by the molecules in the deeper portions of the liquid itself, than they are being drawn outward by the gaseous molecules of the air or vapor outside. In the interior of any liquid, after a very short distance below the surface is reached, each and every molecule is pulled on every side equally, so that a perfect balance of pulls exists. This makes it possible for the slightest disturbance to make these molecules move from place to place. But on the surface the molecules all behave as if they were being drawn into the liquid by a stretched rubber band or spring, and the strength of this pull is really something very great. It is because the pull toward the interior of a liquid is so strong that such large amounts of work are required to be done to convert a liquid into a gas, as in the case of changing water into steam. You set a basin of water on the stove, and work is being done through its bottom, which tends to drive, and does drive, the water molecules out through the surface film, but against a pressure which is enormous. Ostwald states that this pressure, for the liquid ether, amounts to something like 1284 atmospheres, or more than 9 tons to the square inch; and yet the latent heat of water is nearly six times as great as that of ether. It is not strange, therefore, that to change a pound of water at 212° F. into steam at the same temperature, and under the pressure of 1 atmosphere, requires an amount of work which, when expressed in the terms of tons lifted against the force of gravity, finds its equivalent in no less than 37 tons lifted 10 feet high.

These enormous pressures are not felt by bodies placed in the interior of water, because they are surrounded by a water surface which pushes away from the body with equal force.

Now in the phenomena of capillarity a small part of these pressures are brought into play, giving rise to the movements of water in the soil, to the ascent of oil in the wick of a lamp, and to other similar phenomena.

If a clean glass tube .055 inches in diameter is placed upright in pure water at a temperature near freezing, the water will be seen to rise on the inside to the height of 1 inch above the surface of the liquid in the vessel, and the smaller the diameter of the tube is the greater will be the height to which the water will rise in it, the height being inversely proportional to the size of the tube at the surface of the water within, as shown in the table below: —

In a tube 1 inch in diameter the water rises .054 inches.
" " .1 " " " " " " .545 "
" " .01 " " " " " " 5.456 "
" " .001 " " " " " " 54.56 "

To understand thoroughly just why water rises to a greater height in the narrow tubes than it does in the wider ones, will help us the more clearly to understand the capillary movements of water in the soil, and as the manner of action of the forces which produce the rise can be more simply stated for the tubes than for the soil itself, it will be best to explain the action as it takes place in the tubes first.

Thrusting a clean glass tube beneath the surface of water and then withdrawing it, it comes forth wet, that is, with a layer of water adhering to it. This proves that the force acting between the molecules of water and those of the glass is stronger than the force acting between the molecules of the water; for, were this not so, the glass must necessarily come from the water dry, just as it does when withdrawn from a dish of mercury.

Not only is the attracting force of the glass molecules stronger than is that of the water molecules, but this force reaches out and is felt over some distance beyond the surface, a distance which Quincke estimates at not far from one five-hundred-thousandth of an inch. Now when the glass tube is thrust into the water, the row of glass molecules just above the water draw up toward them all around a row of water molecules; but as these move up, other water molecules are made to follow, and the whole surface film is somewhat raised. When this lifting process is once started, it goes on until so much water has been lifted above the level of the water in the vessel as will balance, by its weight, the total pull of the molecules in the wall of the glass tube at the margin of the surface film.

Now, as the circumferences of cylindrical tubes increase in length in the same ratio as their diameters do, it is plain that if we take the distance around a tube whose diameter is .001 of an inch as 1, then in the case of tubes whose diameters are .01, .1, and 1 inch respectively, their circumferences will be 10, 100, and 1000 times as long. Their walls will also contain 10, 100, and 1000 times as many molecules, and as each and every molecule is able to do a like amount of work, it follows that the total work done by the walls of these several tubes in lifting water will stand in the relation of 1 to 10, to 100, to 1000.

But the areas of the cross-sections of these tubes, and their volumes also, grow larger in the ratio of the squares of their diameters, so that the surfaces of the columns of water in the several tubes will be in the ratio of 1, 100, 10,000, and 1,000,000, where their diameters are 1, 10, 100, and 1000 respectively.

We have seen before that the height to which the surface of the water is raised is the greater the smaller the diameter of the tube, so that if we call the height to which the largest tube raises its water 1, then the heights in the other cases will be 10, 100, and 1000. But only the very surface layer of water is raised through these heights, while the bottom layer has not been lifted, and, this being true, the average height through which the water has been raised in each case will be .5, 5, 50, and 500.

Now to get a relative measure of the amount of work done in each case, we must multiply the mean height of the respective columns by their areas, and, doing so, we have,—

	Relative Mean Height.	Area.	Total Work Done.	Relative Work Done.
For the 1 inch tube	.5	1000000	500000	1000
" " .1 " "	5.0	10000	50000	100
" " .01 " "	50.0	100	5000	10
" " .001 " "	500.0	1	500	1

It is here seen that while the amount of water lifted, or the work done, by the walls of the largest tube is much more than that done by the smallest tube, it is only so much more as the number of molecules constituting its circumference is greater. It is also seen that before the molecules in the walls of the smaller tubes have accomplished relatively the same amount of work as those in the walls of the larger tubes, they must necessarily have lifted their columns of water higher, and because the weight of the water in the column increases as the square of the diameters of the tubes, but only directly as the heights.

In order to apply these principles to the capillary movement of water in the soil, we need to call to mind the cavities or openings which are left between the soil grains, which are similar to, but much more irregular, than those left between shot when they are filled into a dish. While these cavities do not form straight capillary tubes leading from deep in the ground to the surface, they do in fact form broken or zigzag passageways which in effect act in the manner of the capillary tubes, but with greater resistance to flow, on account of the water being forced, during the course of its rise, to change its direction so many times before reaching the surface of the ground or the root of a plant.

If we ask ourselves how much work this surface tension, or capillary power, is able to do, the answer must be sought in the amount of water which can be raised by it through a known height in a given time. The writer has found that a very fine sand did lift water through four feet at the rate of .91 pounds per square foot in twenty-four hours, while a clay loam lifted it at the rate of .9 pounds per square foot through the same distance in the same time. Now, this amount of work, when measured in horse power of 550 pounds lifted one foot high per second, is very small; so small indeed is it, that one horse power may do the effective work of the surface tension for 302 acres when moving water at the observed rate mentioned above.

It would be a very grave mistake, however, to conclude that because this form of energy is so small in amount, we may pay no heed to it. On the contrary, it is because the amount of work it can do is so small that it becomes of the utmost importance to so manage the soil as to enable it to act under the most favorable conditions. And when it is stated that this surface tension is the only available power for moving the water of the soil to the roots of plants, then there can be no question regarding the importance of all those processes which tend to favor capillary movement and the conservation of soil moisture.

THE NATURE OF SOLUTION.

Let us now consider the process of solution, the means by which the soil gives to the capillary water the ash ingredients and the nitrogen compounds of plant food. The real nature of this process will be most easily understood if we first consider what takes place when we detect the odor of some fragrant flower or the perfume from some unstoppled bottle placed at a distance from us. In these cases there are traveling away from the blossom, or from the open bottle, molecules of that substance whose odor we detect; these perfumes are going into solution in the air, and we detect them as they are brought into contact with the organs of smell. We know there is a movement of the perfume in these cases, because they reach us at a distance; we know, too, that this movement is taking place in all directions, and at the same time, because, on whatever side we stand, the odor comes to us. We know, too, that these molecules tend to spread wider and wider, until the whole air has been permeated with them. The evaporation of water is simply another case of solution in the air, where the water molecules, through the absorption of heat, have been so violently disturbed as to be thrown beyond the sphere of molecular attraction in the liquid, when they become a gas and travel indiscriminately in all directions through the air.

When a lump of salt or of sugar is dropped to the bottom of a vessel of water, our experience has demonstrated time and again that, after a longer or a shorter interval, and in spite of the fact that these substances are heavier than water, they will be found uniformly distributed throughout the whole. Here, too, the molecules of sugar or of salt wander about from place to place in the water, until there is no portion of it they have not reached. At the outset, the action of surface tension results in overcoming the cohesive or binding power which makes the sugar a solid. Its molecules are being dislodged, and, once free to move, they are never again relatively at rest until some cause is brought into operation to convert them once more into a solid.

So, too, when water is brought into contact with the surface of soil grains, or particles of various kinds of fertilizers which may chance to be present among them, there is here set up a disengagement of the surface molecules, and these, finding themselves free in the capillary water, travel from place to place in it, or they are borne along with the current movements as they rise to meet the losses due to evaporation, and pass laterally to make good the amounts withdrawn by the roots of plants.

Nearly all substances dissolve more rapidly at high temperatures than at low, so that a warm soil prepares food faster than a cold one can, and this brings us to consider the work which altered sunshine does in the preparation and movement of plant food in the soil. In speaking of the surface tension of liquids, we have referred to the strong power which is exerted among the molecules, tending to draw them together, and which is only overcome through the expenditure of an enormous amount of heat energy when evaporation takes place. Now, as a basis for understanding the solution of plant food in soil water, let us get clearly in mind that, fundamentally, the solution of a solid in water is not very different from the evaporation of water in air, and let us keep in mind, too, that, strong as the attractive force is between molecule and molecule of water, the dynamic effects of heat are great enough, even when water is frozen into solid ice, to throw its molecules entirely beyond the bounds of cohesive attraction, thus causing frozen garments to dry rapidly even when hung in an atmosphere far below freezing. Solid camphor evaporates, too, as we know from its odor, under the dynamic action of heat.

We have seen that it is only the *surface* molecules of water which find great difficulty in moving from place to place. Within the liquid mass the forces are very great, but so nearly balanced on all sides that only a little heat energy is required to cause a movement. So, too, in the case of the soil grains, the surface molecules are under a strong tension, which makes it extremely difficult for altered sunshine to throw any of them out, or, in other words, to cause them to evaporate.

But when a soil grain becomes invested by a film of water, this water, through its strong outward action, so much weakens the surface tension of the soil grain that the absorbed heat, the altered sunshine, is now able to throw some of its molecules through the surface film and into the water outside, thus dissolving it, and this process of solution will go on until the water which surrounds the soil grain has become saturated with the dissolved substance.

Through the studies of van't Hoff, W. Nernst and others, it appears that water saturated with a dissolved solid is not widely different from air saturated with water; and it is now generally agreed that a saturated air is simply one so full of flying molecules of water that just as many of them fall back again into the water each second of time as are driven out of it through the action of heat. In this view evaporation has not stopped, but rather the rate of condensation has come to equal the rate of evaporation. If, however, the temperature should fall, then the stream of outflowing molecules would become smaller, because the engine is slowing down, while the incoming molecules, reaching the water, remain there, thus rendering the air drier. On the other hand if, when the air has become saturated, the temperature of both air and water should rise,—that is, if altered sunshine should enter the water at a more rapid rate,—then more water molecules would be forced out each second than, for the time being, are returned to it, and hence evaporation would commence again and the air would come to contain a higher per cent of moisture. It would be a stronger solution.

L

If we carry this method of work to the solution of plant food in the soil, we shall understand that, as the ground warms up in the spring, as the altered sunshine becomes more concentrated in the soil grains, as the molecular swing becomes stronger and stronger, the rate at which the molecules of plant food are driven out into the film of water mounts higher and higher until finally the soil water, like the atmosphere, has become saturated. In this condition the play back upon the soil grains is equaled by the outward discharge, and a balance is reached, but not a condition of rest; for solution and precipitation are going on even-handed. If the temperature of the soil should go up, then the amount of plant food in solution would increase; while if it should fall, then some of the dissolved materials would again take on the solid form.

On the other hand, if the root hairs of plants, or the cell walls of the microscopic life of the soil, come in contact with the water in which this process of solution is being carried on, and they utilize a portion of these dissolved materials, then they will cause the solution to go forward at a more rapid rate. The solution will be more rapid because every outgoing molecule which is captured and prevented from returning, with its heavy blow, by so much reduces the resistance which the outflowing molecules are obliged to overcome.

We must not lose sight of the fact, in this connection, that the capillary movement which bears the dissolved food materials toward the surface of the ground and to the roots of plants must in some way represent altered sunshine. Somehow the molecular swing of the soil grains is transformed into current flow; for a stream of water cannot rise continuously through the soil, the whole summer through, against gravity, against the friction of the very great soil surface, and against the resistance necessitated by the many turns the currents are forced to take, without the expenditure of much more energy than is represented by the amount of water lifted.

OSMOSIS AND DIFFUSION IN PLANT FEEDING.

We have seen how altered sunshine, working through the soil moisture, brings into liquid form the food of plants; we have seen, too, how this prepared food is borne along in the capillary currents to the places where it is to be used. Let us next consider how the food-laden water, after reaching the root hairs, is made to enter them and rise through the stems into the leaves and sunshine above.

The process by which this work has been accomplished has been technically named osmosis, and we have been made perfectly familiar with some of its results in many ways. When dry beans, prunes, or raisins are put in water to soak, it is not long before they have increased in size through the absorption of water. Osmosis has taken place in each of these cases, and they illustrate, so far as the real mode of action is concerned, the flow of sap in plants. When juicy fruits are covered with sugar or are placed in a strong syrup, they shrivel and become much smaller; here, too, osmosis has occurred, but the strongest movements have taken place in the opposite direction. In both of these classes of cases movements have taken place, very similar to evaporation and to solution, and the dynamic power of heat or the energy of molecular motion has been the actuating energy.

In trying to understand this process as it comes into play in the feeding of plants, we need first to consider some physical experiments which have served to demonstrate the intensity of pressures which, under favorable conditions, are developed during the action of osmosis. The swelling of the dry bean to two or three times its original volume certainly represents no small amount of pressure; for it is evident enough that, could we connect a column of water with the bean and undertake to distend it by direct hydrostatic pressure, a very high column of water would be required to do the work.

Abbé Nollet, who lived between 1700 and 1770, appears to have been the first to record that, if a glass vessel be filled with wine and covered with a bladder and then immersed in water, the contents of the vessel would increase and sometimes to such an extent as to burst the membrane. The same fact was rediscovered a second, third, and even a fourth time before it came to be the common property of science and was thoroughly investigated.

It was reserved, however, for W. Pfeffer, in 1877, to conduct experiments in osmosis under such conditions that a full measure of its power could be obtained. He used a porous clay cell, lined with a membrane, produced by precipitating on its inner wall a coating of copper ferrocyanide. Attaching a pressure gauge to this, he found that when this cell was filled with a solution containing 1.5 per cent of potassium nitrate and then set in a vessel of pure water, the water continued to flow into it until a pressure of 3 atmospheres had been reached, and this is equivalent to a column of water at sea level, more than 100 feet high; while de Vries showed that a solution of potassium citrate developed a pressure of more than 5 atmospheres, equal to a column of water rising to a height of 170 feet, or a pressure exceeding 70 pounds to the square inch. It is not strange, therefore, that dry beans swell when placed in water, nor that dry wood, when becoming wet, exerts such high pressure.

When two dissolved or liquid substances are separated by a membrane through which one of these liquids can pass more readily than the other, then that liquid whose molecules are small enough to enable them to make their way through the separating membrane most rapidly, does so. This causes an accumulation, within the bean, for example, thus increasing its volume and at the same time the pressure. Under these conditions the molecules of water wandering about or diffusing, impelled by their heat energy, continue to pass through the porous membrane, continue to pass into the bean, until the same number, in their vagrant wandering, tend to emerge from the porous membrane, from the bean, each second of time, as tend to enter it. When this state has been reached, the in-and-out play goes on even-handed, a balance has been reached, and apparent change has stopped; a condition similar to air saturated with water has been attained.

We may consider first the movement of water in and through the plant. If in any plant cell water is being consumed as food, if it is being changed into some other substance, or if it is evaporating and leaving the cell in this way, the removal of these molecules from the solution leaves space unoccupied, diminishes the osmotic pressure at that place, and this diminution of pressure causes a flow of molecules from contiguous cells to make good this loss. Suppose that in a certain portion of a plant water is being used in the manufacture of sugar or starch. Then so long as this demand continues, there will be a lowering of the osmotic water pressure, and so long will water move toward that place. It is as if there were a long line of vessels full of water, all of them connected by means of pipes, and at one end of the line water is being dipped out, so that the water level or pressure in that is being lowered. When this is done, water will flow from the nearest vessel to take the place of that which has been removed, but no sooner is the water pressure or level in the second vessel lowered than a flow sets in from the third, and this to be followed by water from the fourth, fifth, and so on to the end of the line. Then, if water is added to the far end of the line as rapidly as it is being removed at the other, a continuous flow will be maintained. On the other hand, if the dipping ceases, then the water surface becomes level throughout the whole system, the water pressure becomes uniform, and flow ceases.

Now in the case of the plant supposed, the diminution of the osmotic water pressure at the place of growth is propagated backward toward the end root hairs in the ground, and so soon as some of the root cells which are in contact with the soil water have their pressure reduced by the flow toward the place where the water is being used, then the balance of pressure between the water in the root hairs and that which is in contact outside is destroyed, and more molecules enter than escape, impelled by the dynamic energy of molecular swing. So, too, if the reduction of osmotic pressure in the leaves is due to evaporation rather than the fixation of water, the same result follows.

Then regarding the movement of nitrates from the soil into and through the plant; so long as there is no consumption of this substance in the tissues, just so many molecules of the nitrate, impelled by altered sunshine, leave the soil water and wander through the fluids of the plant as are required to produce a balance of osmotic pressure throughout it, or, what is the same thing, an even distribution of nitrate in the sap. Under these conditions accumulation ceases. If, however, growth is going on which results in the removal and fixation of the nitrate, then, where this is taking place, the back play of the molecules is reduced, the balance is destroyed, and more nitrate advances to make good the loss; a flow sets in from the soil toward the place of consumption.

It follows from what has been said regarding osmotic pressure and its action, that there is a tendency for each ingredient of plant food to travel in a large measure independently of every other material, the movements not being in the nature of currents. So that if the escape of water from the plant were to be stopped altogether, the osmotic pressure might be fully able to convey nitrogen and the ash ingredients to the plant dissolved in the water which, under these conditions, would act simply as a medium through which the diffusion is carried on.

It will also be understood from these statements that the transfer of the assimilated products to the places where they are stored, whether in the seeds, thickened leaves, or fleshy roots and tubers, must also be effected by this process of diffusion through osmotic pressure. As the soluble forms, out of which the insoluble starches, gums, and fats are constructed, are withdrawn from solution, at their respective places, there must necessarily result a diminished pressure in the ear of corn, in the head of wheat, in the expanding apple and in the tubers of the potato, the bulb of the onion or the root of the carrot, toward which the products compounded in the green parts of the plants must be impelled by the energy of altered sunshine.

In justice to the reader and to the truth, it should be said in closing this subject that not all physiologists are fully agreed that osmotic pressure is the only power which actuates the movement of sap in plants, but the most reliable data we now have go to show that it is the prime, if not the sole, mover, and to this physiologists are agreed.

The so-called selective power of plants, whereby they obtain those substances dissolved in the soil water which contribute to growth, and those substances which are brought into solution by the corroding power of juices exuded from their roots, finds its explanation in osmotic pressure. Regarding this point, it should be understood that any substance held in solution in the soil water may make its way through the tissues of plants, and usually does so until the sap contains the same number of molecules per unit volume as the soil water contains. Under these conditions, if that substance is not being used by the plant, no further accumulation of it can take place; for the outflow from the root into the soil keeps exact pace with the inflow from the soil toward the tissues. The loss of water by evaporation through the surface of the plant or the consumption of it as food, which tends to make the strength of the solution of those substances not used as food stronger, cannot result in a permanent increase of them in the plant, because, unless these substances are actually taken out of solution, they travel back toward the root again and escape into the soil water so long as the solution inside is stronger than is that outside.

If the molecules of any substance in solution are too

large to easily make their way through the walls of root hairs, as seems to be the case with some of the colloidal salts, then there is a positive exclusion of these substances from the plant, or they only enter in diminished quantities, just as the colloidal substances in the plant cells do not readily escape into the soil water under the impulse of this osmotic pressure.

So, too, when a soil contains any substance which is poisonous to the plant or which by its presence in the sap would be harmful, the roots have no power to exclude it. The process of osmosis tends to carry it to all parts of the plant, just as when a soluble poison is taken into the stomach of an animal it is absorbed and goes quickly through the system.

The plant then simply takes out of the solution brought to it those substances which it needs. It is not overloaded with other substances, because it leaves those in solution, and this is a sufficient check to a further accumulation of them.

CHAPTER V.

SOIL WATER.

From what has been said in the preceding chapter, it is evident that water plays an extremely important part in the fertility of soils and in the feeding and life of the plant. Surrounding the soil grains as a surface film, it acts directly upon even the most difficultly soluble of the soil ingredients, taking them into solution in larger or smaller quantities. Holding carbonic, humic, and other acids in solution and bringing them into close contact with the surfaces of the soil grains under a diminished surface tension, these substances are enabled to bring into soluble form the ash ingredients of plant food as they could not without the aid of water. Wood kept permanently so dry that it contains only a small amount of hygroscopic moisture never decays, and such perishable articles as fruits and even flesh may be preserved for indefinite periods of time if only they are kept permanently dry; and so in the return of dead organic matter in the soil to the ash ingredients, nitrates and carbon dioxide, the forms in which they can be used over again by higher plants, the presence of a certain amount of soil moisture is indispensable.

After the plant food has been prepared in the soil or in the air, it is useless until endowed with the possibilities of movement toward and through the living tissues. But water, through the action of capillarity and osmotic pressure, is the medium of transport by which the ash ingredients and the nitrogen of the soil are moved to the roots of plants, by which they are drifted into the sunshine of the laboratories in the green leaves and bark, and from which they are again taken to their final places in the structure of the plant. Nor is this all, for water is itself a food substance used in large quantities by all plants of whatever sort. By its evaporation from the foliage of plants, it not only holds the temperature down within the normal range of the vital processes there going on, but, because of this lowering of the temperature, it also hastens the osmotic flow of sap toward the leaves.

AMOUNT OF WATER USED BY CROPS.

The amount of water demanded by crops under our methods of culture is very large. So large, indeed, that in Wisconsin the following amounts in tons of water per ton of dry matter and in inches of water per ton have been lost by transpiration through the plant and evaporation from the soil:—

Dent corn used 309.8 T. equal to 2.64 in. of water per 1 T. of dry matter.
Flint   "   "   233.9 "   "   2.14 "   "   "   "   1 "   "   "   "
Red clover "   452.8 "   "   4.03 "   "   "   "   1 "   "   "   "
Barley used   392.9 "   "   3.43 "   "   "   "   1 "   "   "   "
Oats   "   522.4 "   "   4.76 "   "   "   "   1 "   "   "   "
Field peas used 477.4 "   "   4.21 "   "   "   "   1 "   "   "   "
Potatoes used 422.7 "   "   3.73 "   "   "   "   1 "   "   "   "

Hellriegel found, through experiments conducted in Prussia, that the amounts of water withdrawn from the soil and given to the air, almost wholly through the plant, were as follows:—

Barley used 310 pounds of water for each pound of dry matter produced, summer rye 353 pounds, oats 376 pounds, summer wheat 338 pounds, horse beans 282 pounds, peas 273 pounds, red clover 310 pounds, and buckwheat 363 pounds of water for one pound of dry matter. This is at the mean rate of 325 tons of water for each ton of dry matter produced.

These large amounts of water used by vegetation in the maintenance of its functions are necessitated partly by the rapid evaporation which results from the extreme exposure of the foliage under conditions best suited to facilitate a rapid loss of water in this way; but evidently also, in no small degree, by the physiological processes of the plant itself, which demand a large movement of water through the growing tissues.

The fact, however, that the growth of vegetation is often extremely rapid during the times when the rate of evaporation must be relatively slow, makes it appear more than probable that the large losses of water by evaporation from cultivated fields, through the transpiration of growing crops, is not really demanded, so far as the needs of growth are concerned; and if this is true, it follows that wherever we can adopt means which tend to avoid needless loss of water through the foliage of plants, we are practicing economy regarding soil moisture just as we are in our methods of mulching, by cultivation or otherwise, which aim to reduce the evaporation from the surface of the soil.

There are very few countries, indeed, where the distribution of rainfall in time and amount is such as to permit fertile soils to produce the largest crops they are able to bear; and this being true, those soils which are able to store the largest quantities of rain in a condition which shall permit vegetation to use it to the best advantage are likely to be the most productive. On this account the water capacity of soils is an important factor in the determination of land values.

CAPACITY OF SOILS TO HOLD WATER.

Since each independent soil grain of a moist soil is more or less completely surrounded by a film of water, it is evident that, other conditions being the same, that soil whose grains present the largest aggregate surface area may retain the most water per cubic foot. Now a cubic foot of marbles one inch in diameter possesses an aggregate surface of 37.7 square feet, while if the marbles were reduced in diameter to one one-thousandth of an inch, then the total area per cubic foot is increased to 37,700 square feet. From these differences it is evident that the amounts of water coarse and fine grained soils retain will be very different.

Under field conditions the amount of water a soil can hold varies, not simply with the size of the soil grains, but also with the distance of standing water in the ground below the surface and the length of time which has elapsed since it has rained.

To illustrate the influence of the size of the soil grains on the amount of water a soil may retain, let me cite some observations upon columns of sand 10 feet long, from which water had been allowed to percolate during 111 days under conditions where no evaporation could take place from their surfaces. The column whose mean grains had diameters of $\frac{186}{10000}$ of an inch contained, in the upper 30 inches, 2.16 per cent, in the second 30 inches, 2.41 per cent, in the third 30 inches, 2.73 per cent, and in the fourth, 7.77 per cent; but a sand whose average grains were $\frac{45}{10000}$ of an inch in diameter retained 3.06 per cent for the upper, 3.71 per cent for the second, 5.46 per cent for the third, and 18.05 per cent for the lower 30 inches. The first column had retained only

3.77 per cent of its dry weight of water, while the second, or finer, one had retained 7.57 per cent of its dry weight, or about twice the amount. Two other sands, with grains about $\frac{73}{10000}$ and $\frac{61}{10000}$ of an inch in diameter, placed under the same conditions, had retained 4.92 and 5.76 per cent respectively; that is to say, columns of sand 10 feet long composed of grains

$\frac{186}{10000}$ of an inch in diameter retained 3.77 per cent of water,
$\frac{73}{10000}$ " " " " 4.92 " "
$\frac{61}{10000}$ " " " " 5.76 " "
$\frac{45}{10000}$ " " " " 7.57 " "

after a period of percolation of 111 days.

It will be understood that the smaller the size of the soil grains, the smaller will be the size of the pores through which the water is obliged to flow in percolating downward under the influence of gravity, and hence the slower will be the rate at which it will move.

The table given below will show the influence of the size of the soil grains on the rate at which the water may be lost by percolation downward. The results here given were obtained from columns of sand 8 feet long, and composed of grains having the same diameters as those just referred to.

At the end of 9 days these several soils had lost water, in the order from the coarsest to the finest, 15.29, 14.35, 12.86, 10.02, and 8.42 per cent respectively of the dry weight of the sand.

A strange thing about these soils is that they continued percolating slowly for more than 259 days, and during the last 250 days they lost respectively, beginning with the coarsest, at the rate of 9.15, 7.77, 6.56, 7.69, and 7.56 pounds of water for each square foot of surface,

RATE OF PERCOLATION FROM EIGHT FEET OF SAND OF DIFFERENT DEGREES OF FINENESS.

Size of Grains.	Time of Percolation.			
	1 Hour.	2 Hours.	24 Hours.	48 Hours.
	Per cent.	Per cent.	Per cent.	Per cent.
$\frac{186}{10000}$ of an inch . . .	9.10	10.45	13.05	13.52
$\frac{73}{10000}$ of an inch . . .	7.95	9.47	12.31	12.72
$\frac{61}{10000}$ of an inch . . .	6.22	9.21	11.71	11.53
$\frac{45}{10000}$ of an inch . . .	1.76	2.83	7.64	8.44
$\frac{32}{10000}$ of an inch . . .	1.28	1.91	5.83	6.79

amounts ranging from 1.2 to 1.8 inches of rain; nor had percolation at this time ceased.

It must be understood that with true soils of much finer texture, the rate of percolation from them would in all probability be much slower and the water-holding power much larger, but no measurements have been made for any of these under the conditions stated above.

To show what actual field soils may hold when their surfaces are only 11 inches above standing water, this water having been lifted into them by capillarity, the following results may be cited: —

	Per cent of Water.	Lbs. of Water.	Inches of Water.
Surface foot of clay loam contained	32.2	23.9	4.59
Second " reddish clay "	23.8	22.2	4.26
Third " " " "	24.5	22.7	4.37
Fourth " clay and sand "	22.6	22.1	4.25
Fifth " fine sand "	17.5	19.6	3.77
Total	. . .	110.5	21.24

Then, again, to show the water-holding power of natural soils under field conditions, the table below gives the water content of a piece of fallow ground at three different times during the season, the soil being a medium clay loam with clay subsoil in the second and third feet and followed with sand in the fourth foot. The field was tile drained, the drains being 4 feet below the surface of the ground.

Depth.	May 16. per cent.	July 13. per cent.	Aug. 30. per cent.
First foot }	25.77	24.27	24.71
Second foot }		23.62	24.30
Third foot }	22.87	23.52	24.03
Fourth foot }		23.32	22.29
Average	24.32	23.68	23.83

It is evident from the cases cited that the water-holding power of soils under field conditions is really large, and that, where there is no crop upon the ground to consume the water, and the water table is not more than 5 feet below the surface, the ordinary summer rainfall, assisted by capillarity, may nearly conpensate for the loss of water by percolation and by surface evaporation; but this is far from being true when a crop is occupying the land, unless the rainfall is more than usually heavy and frequent.

It may be said in general, regarding the capacity of soils to hold water, that the finer the soil grains the more the soil will hold, and the greater the number of spaces which are larger than capillary size the less it will hold. The capacity of a soil for water decreases to some extent also as its temperature rises, but under field conditions the effects of temperature are not large. It is certainly true for sandy soils that their capacity for water decreases in a marked manner the higher their surfaces are above standing water, and it is probably true that all soils, unless we must except the finest clays, fall under the same law.

If it were true that all of the water which a soil may retain were available to crops, and it were not necessary to allow some of the water of completely saturated soils to percolate away before agricultural plants will thrive in them, then in most humid regions it would be possible to obtain much larger crops than can now be raised without irrigation. The facts are that not only must a considerable amount of the soil moisture be left unused by the crop, if large yields are expected, but from 30 to 40 per cent of their saturation amounts must have drained away before the soil can contain air enough to maintain the breathing of ordinary roots and germinating seeds. Hellriegel places the amount of water which should drain away from the soil before it becomes habitable by cultivated plants as high as 40 to 50 per cent of their full capacity, and in order that maximum results may be reached, the water content should not fall very far below these amounts.

The soils having a small water capacity will yield their water to plants more readily and completely than the heavier types. The writer has observed that, in a sandy soil whose maximum water capacity was about 18 per cent, corn was able to draw the water in it down to 4.17 per cent, while in a clay soil, having a water capacity of about 26 per cent and lying nearer the surface and farther from standing water in the ground, the same plant had only succeeded in using the water down to 11.79 per cent.

Now if we compare the absolute amounts of water

given over to the corn by these soils, we shall find that the sandy soil contributed 13.83 pounds of its amount per cubic foot, while the clay had only yielded 12.5 pounds. It thus becomes evident that while the percentage capacity of the sandy soil is much below that of the clay, its greater weight per cubic foot and the greater freedom with which it yields water to plants makes its storage capacity of available water more nearly equal to that of the loamy clay than would at first be supposed. It is on this account, in part, that a sandy soil, kept well fertilized, has many advantages over the colder, less perfectly aërated, and more obstinate clayey ones, which crack badly in excessively dry weather and become oversaturated in wet seasons.

It follows from what has been said regarding the capacity of long columns of soil for water, that the distance of standing water below the surface, or below the level to which the roots of cultivated crops may reach, must materially influence their agricultural value. In general the nearer the surface of the ground water or water table is maintained to the lower surface of the root zone, the more productive the lands will be; for then capillarity is able to hold the water content of the soil more nearly up to the standard required for maximum yields. This is the chief reason why lands which require underdraining are so valuable when they have been thus improved.

THE GROUND WATER AND WELLS.

In all humid climates where the soil, or unsolidified sand and gravel, has a depth of 50 to 200 feet, there is a certain distance below the surface at which all of the interspaces between the incoherent grains are completely filled with water, and the upper surface of this water-filled soil is called the water table. Above this water table the soil contains only capillary water. When wells are sunk into the water-filled soil, the surface of the water in these wells represents the level of the water table, at that place. We should think of all permanent lakes and ponds, too, as extensions of the water table, where the surfaces of low lands are beneath it. It must not be understood, however, that the height of the water table under the land is at the level of the water in the lakes. On the contrary, the water table almost invariably rises as the distance from the margins of lakes and other permanent bodies of water increases. Indeed, the surface of the water table follows in a rough way the general contour of the land, the water-filled soil beneath the surface standing highest where the ground is highest, and lowest where the land is low. The water table has its hills and valleys, which coincide with those of the country it underlies, only they are not as steep, the water being nearer the surface under the low ground than it is under the high. An inspection of Figs. 21 and 22 will show how the ground water is related to the surface in one locality. In one of these figures the contour lines show the surface of the land, while in the other they show the surface of the water-filled soil beneath, on the date there specified, as determined by measurements taken at the wells indicated by the numbers in the figure.

There is one marked difference between the hills and valleys of the water-filled soil and those of the surface of the ground which overlie them, and that is that the height of the water table is not constant. This surface

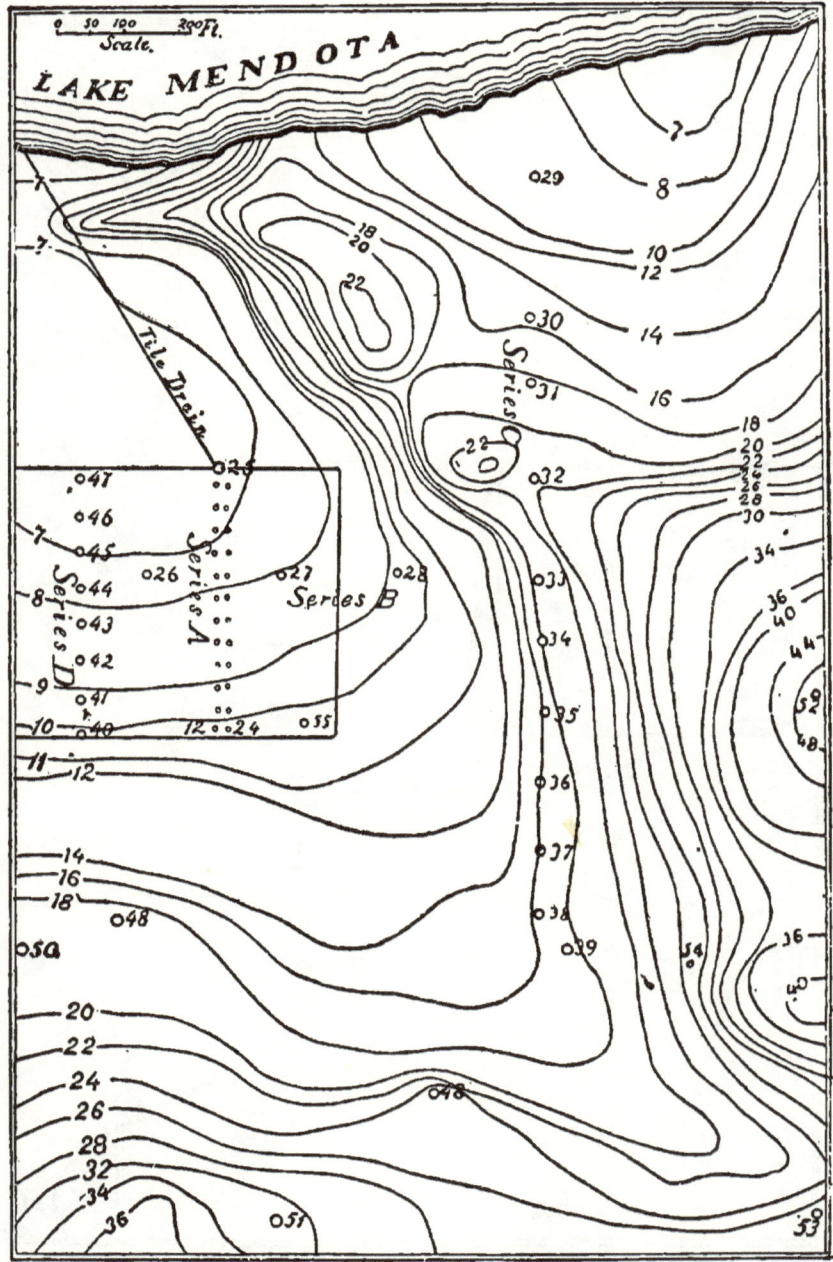

Fig. 21. — Showing the contour of the surface of the ground. Figures in the lines show height of contours above lake level; other figures indicate wells where the height of the water table was measured.

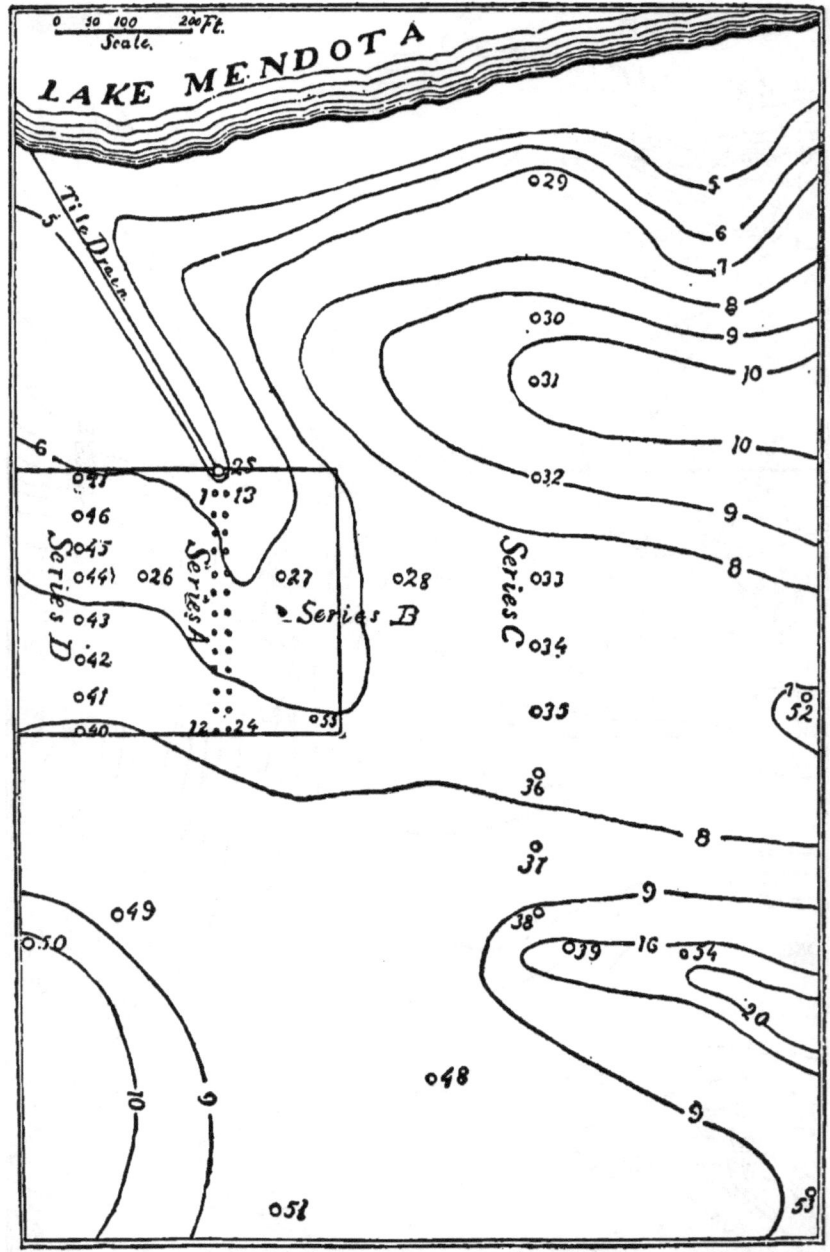

FIG. 22.—Showing the contour of the water table beneath the surface of the area represented by Fig. 21.

rises and falls with the season, being higher usually in the early summer and lowest late in the winter. Its surface, too, in many places, tends to rise higher and higher when several wet seasons succeed one another, and then to fall with the recurrence of dry years.

The reason why the surface of the water-filled soil is not level is because the soil grains offer so much resistance to the lateral flow of the water toward lakes and streams, that the rains which percolate beneath the surface do not have time to drain away before another rain comes to add its water to the soil. But if there should be a long term of years with little rainfall, the water table would sink lower and lower, and fastest where it is highest, until it became nearly or quite horizontal.

When wells are sunk into the water-filled soil, it is evident that water may be reached at very different levels in different localities, and that it will not be necessary to dig to the level of some lake or stream before water is reached, as is commonly believed. To illustrate how far this idea is from being true, it may be cited that a well on the campus of the University of Wisconsin has its water surface 52 feet above the level of Lake Mendota, only 1250 feet distant, and yet this well is dug all the way in a coarse sand and gravel which makes up the axis of an isolated glacial drumlin from 88 to 109 feet above the surface of the lake. Now the water of this well owes its origin wholly to the rain which falls upon the hill and percolates slowly into it; and how slowly this may be can be understood from what has been cited regarding the percolation from the columns of sand only 8 to 10 feet long.

In digging wells into the water-filled soil, it is evident that those sunk in the valleys will be less affected by seasonal changes and will usually give a larger water supply, for the reason that the water under the higher lands tends to flow toward the valleys below ground just as it does above ground. Then, too, in digging wells, care should be taken to sink the bottom of the well so far below the level of the water table, that seasonal changes will not cause it to go dry. More than this, the larger the supply of water which must be drawn from the well, the deeper it should be sunk below the level of the water table, and if the well is being dug during the season when its natural level is highest, and particularly if a series of wet years have preceded the digging, then a special effort should be made to carry the bottom a considerable distance below the level at which water will stand in it at the time. The deeper the well is below the water-filled soil, the farther the water surface may be lowered below the level of the water table, and the farther the pumping carries the water level in the well below that of the water-filled soil, the faster will the water flow into the well, and the larger will be the amount of water it can deliver in a given time.

There is another matter regarding such wells which should be emphasized here, and that is the danger of surface waters percolating into them and rendering them unfit for use. The soil containing only capillary water above the water-filled zone has in its interspaces larger or smaller quantities of air, so much so that all space not occupied with water is taken possession of by it. Now when heavy rains come, the surface soil is first completely filled with water, and under these conditions it is at first difficult for the water to enter the soil deeply and also for the now confined soil air to escape. Then the air, borne down by the pressure of the water above, has a tendency to move away in any direction where it can

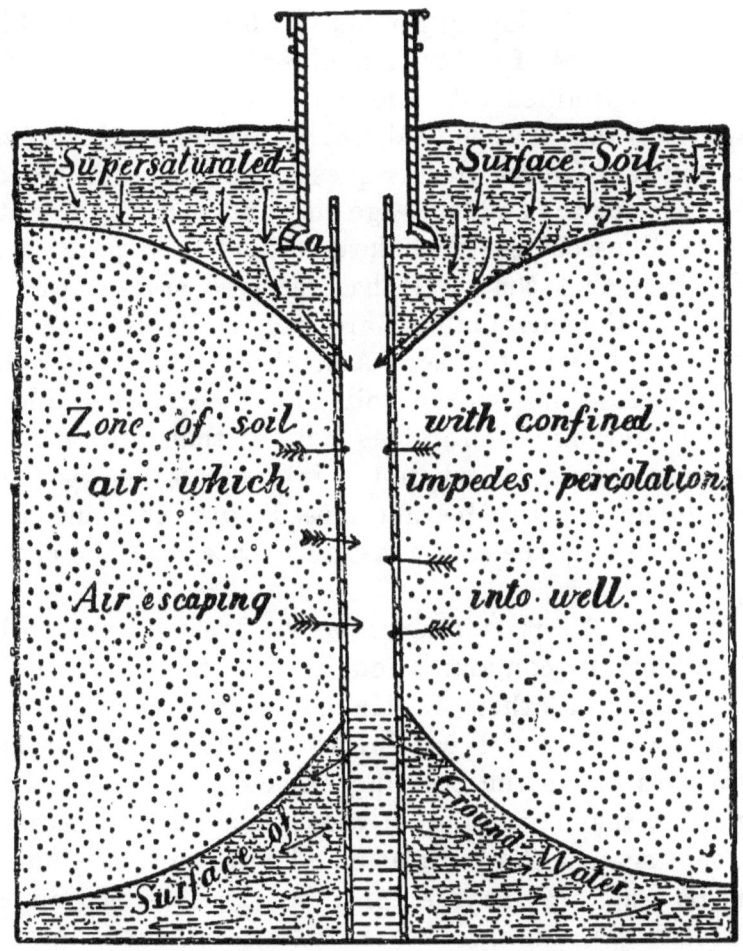

Fig. 23. — Showing the method of percolation of water into and out of wells.

find the easiest escape, and whenever the well is not provided with a water-tight curbing, the air tends to escape into the well, and then the percolating waters follow it under the conditions represented in Fig. 23. In consequence of this tendency of both the air and the water to flow toward and into the well, the confined soil air forms a sort of broad, sloping funnel, down the sides of which large quantities of water from close to the surface of the ground are drained into the well, carrying many of the impurities they may have dissolved in the neighborhood. The sanitary aspect of this question should not be lost sight of; for the sudden large rise and fall of the water in a well associated with heavy rains is a pretty sure indication that that well has received surface waters, which, if the surroundings are bad, is liable to render the water unsafe to use. And observations show that wells in close, impervious soils are much more liable to surface contamination in this manner than are those in the more open soils, where the interspaces near the surface do not readily become closed with the fine sediment moved by the water, as is the case with the clayey types.

Just how far it is practicable to protect wells which are subject to contamination in this manner by using iron tubing or other similar impervious curbing, is a matter which merits careful investigation; for it is a vital question in the building of country homes. It is generally taken for granted that wells thus constructed are safe against the infiltration of surface water, and it may be true to a large extent; but it does not appear improbable, in pumping water from a well tubed up with iron, that the rapid withdrawal of water from about the immediate terminus of the tubing would tend quite as strongly to bring down a new supply from the supersaturated soil above, as to induce it to come from below or from lateral directions, and, if this is true, it is evident that the surface surroundings of a well used for domestic purposes should be scrupulously cared for, even when provided with impervious curbing.

When we consider the movements to which the soil water is subject, it must be said that these belong to not less than three classes, — those due to the action of gravity upon the soil water direct, those due to capillarity or surface tension, and again those which result from changes in the tension or pressure of the soil air.

PERCOLATION MOVEMENTS OF SOIL WATER.

The movements of the gravitational class are the most extensive, the most rapid, and find an expression of their aggregate magnitude in the vast volumes of water which the great rivers of the world pour into the sea. To appreciate the full significance of these movements of the ground water, we need to think of the water-saturated soil as being almost as extended and as continuous as the land areas of the world; for, beneath almost every square foot of soil-covered surface, unless the frozen zones and desert regions must be excepted, there is a perennial percolation of water, at first downward in the slow fashion already referred to as taking place in the sand, but as the deeper, and often more porous, layers are reached, the movements become stronger, and finally the waters emerge as springs, at the foot of hillsides or in the bottom of lakes.

The direction of percolation is by no means always downward nor even lateral. In many places it comes to be vertically upward, toward the surface, giving rise to larger or smaller areas, which require underdraining before they are fit for agricultural purposes. In very many localities, where a sandy or otherwise porous stratum underlies both the flat, marshy areas and the surrounding higher ground, the much greater elevation of the water table under the high ground causes it to act by direct hydrostatic pressure, to force the water to the surface on the low ground, not in the form of springs, but in a broad sheet which tends to keep a wide area more or less uniformly wet. Such areas, when they are underdrained, come to be the most productive lands we have, and largely because of the natural sub-irrigation which is, in such cases, taking place; and it must be remembered that lands so situated are receiving, with the underflow of water, not simply the lime, in the form of carbonates, which may have been dissolved, but also some of the nitrates and other soluble forms of plant food, which may have percolated beyond the reach of root action on the higher ground.

The rate at which water percolates through soils varies between wide limits. In a coarse, sandy soil, whose grains pass a screen of 40 meshes to the inch, but are retained by one of 60, the writer found that water would percolate through a column 14 inches in length and .1 of a square foot in section, at the rate of 301 inches in depth in 24 hours; where the grains passed a screen of 60 meshes, but were retained by one of 80, the percolation was 160 inches in the same time. Reducing the size of the grains still further, so as to pass a screen of 80 meshes, but not one of 100, then the percolation falls to 73.2 inches in 24 hours. While sand, which passed a screen of 100 meshes to the inch, allowed 39.7 inches to pass. A clay loam, on the other hand, was so much more impervious that only 1.6 inches passed a column of the same length in 24 hours, while a fine-textured black marsh soil passed only .7 inches in the same interval.

Adding 20 per cent of blue clay to the sand which percolated 73.2 inches during 24 hours reduced the amount passed through it to only 10.5 inches, while 50 per cent of clay caused a decrease to 2.7 inches during the same time. In all of these cases, while percolation was going on, the several soils were kept covered with water to a depth of 2 inches.

Wollny showed, from experimental studies, that, under like conditions of pressure, soils having different sizes of grains allowed water to pass through columns 11.81 inches deep and 1.97 inches in diameter, at the rates indicated below:—

						Depth in Inches.
Quartz	sand	.0004 to .0028 in. in diam.	percolated	7.27 in 24 hours.
"	"	.0028 to .0045 " " "	"	68.59 " " "
"	"	.0045 to .0067 " " "	"	669.7 " " "
Calcareous	"	.0004 to .0028 " " "	"	6.64 " " "
"	"	.0028 to .0045 " " "	"	63.74 " " "
"	"	.0045 to .0067 " " "	"	304.3 " " "

In these experiments of Wollny's, as in those of the writer, the sizes were obtained by sorting the grains with sieves; but it should be observed that this method of separation is not very exact, as it permits varying amounts of grains of smaller sizes to remain among those having the dimensions sought, and it may be this fact which causes the observed differences between the quartz and calcareous sands.

It must be evident therefore, from the facts here given, that soils differ very widely in their tendency to retain the rains which fall upon them, and also in their tendency to lose their soluble plant foods during wet periods. It is evident also that, while a small per cent of clay added to a sand will greatly increase its power to retain water, relatively large amounts of sand would have to be added to the stiff clays to render them very open.

A very wrong impression would be left if it were not stated here that the amounts of percolation recorded above for the several cases are at maximum rates, and that such rates of percolation seldom occur in the field anywhere in nature. Indeed, such rates can only take place after all air has been expelled from the soil and the spaces between the grains have been completely filled with water. This condition rarely occurs except in the spring and after protracted and very heavy rains. After vegetation is once well under way, the surface two and three feet of soil are so rapidly dried that ordinary rains can only partially maintain such slow percolation as has been shown to take place from the long columns of sand cited above. Indeed, it more frequently happens in our climate, during the middle and later part of the growing season, that the rains we do have tend rather to strengthen the upward flow of water toward the surface by capillarity than to result in gravitational losses downward by percolation. Much the larger part of the water which goes to feed perennial springs is that which finds its way into the soil when vegetation is dormant or only feebly active.

CAPILLARY MOVEMENTS OF SOIL WATER.

The capillary movement of water in field soils is the slow creeping over the soil grains due to the action of surface tension, as already described. It takes place in any and all directions, but usually from below toward the surface of the ground where evaporation is taking place, and toward absorbing root hairs where the surfaces of soil grains are being depleted of their capillary water.

The rate at which water can be moved through soils by capillary action is never very rapid when compared with the gravitational movements we have just considered. It is, however, sufficiently great and extensive to be of the highest agricultural importance.

When a cylinder of very fine sand, 12 inches in diameter and 4 feet high, was saturated with water and arranged so that water could be automatically admitted to it from below as rapidly as it could be removed by evaporation from the surface, it was found, on placing this cylinder in a strong current of dry air, and under conditions where the water must be lifted 12 inches by capillarity, that the work of lifting was done at the mean rate of 2.37 pounds per square foot daily during 10 days. On changing the level of standing water in the cylinder to 2 feet below the surface, the mean daily capillary rise came to be 2.07 pounds. When the capillary lifting went on through 3 feet, the work done daily was at the rate of 1.23 pounds per each square foot of surface; while at 4 feet it became .91 pounds.

A similar trial with a medium clay loam gave mean rates of lifting water amounting to 2.05 pounds, where the lifting was through 1 foot, 1.62 pounds where it was through 2 feet, 1 pound where it was through 3 feet, and .9 pound where the rise was through 4 feet of this soil. It is very certain that the rate of capillary rise of water in each of these cases might have been larger had the evaporation from the surface been greater; for the surface of the soil remained wet throughout all of the experiments. It is also certain that the decrease with the depth is larger than it would have been had not the deposit of salts on the surface, tending to form a crust, diminished the evaporation. This fact was proven conclusively in both cases by removing the crust formed on the surface at the close of the trials, when the mean rate of evaporation, and hence also of capillary rise of water through these soils, became 1.38 pounds for the fine sand and 1.27 pounds for the clay loam per square foot and per day.

In another experiment, under perfectly natural field conditions, it was found that the surface four feet of soil lost, during 7 days, 9.13 pounds of water from each square foot of surface, and this, too, under conditions which make it certain that this loss was wholly due to a capillary rise and to evaporation from the surface. This loss gives a mean capillary movement amounting to 1.3 pounds daily. In this field case the water table was 4.5 to 5 feet below the surface, or rather more than in the laboratory trials referred to above.

The writer has also made other field determinations of the capillary rise of water in field soils under field conditions, and has found, as the mean of 10 determinations in as many different places, a measured loss of water amounting to 1.65 pounds per square foot per day. In these trials it is quite possible that a portion of the observed loss may have been due to percolation downward. But, on the other hand, it should be added that it is more than probable that the real losses from the ground upward, and hence also the total capillary work done, were larger than the measured losses, because it is only fair to assume that capillary action was bringing water from below into the layers of soil, where the losses were measured, and if this was true, such action could only tend to make the observed changes less than they in reality were.

It must not be inferred that the capillary action is uniformly as rapid as the cases just cited. These should be looked upon, not as extremes, but as representing rates of movement of the soil moisture toward the surface which are above the average.

When the soil, for any reason, has become very dry, and also when it has been rendered loose and open so that it contains numerous large spaces which are many times the diameters of the soil grains, the rate of capillary rise of water then becomes much slower. In illustration of the slower rise of capillary water in a dry soil may be cited the case where cylinders of dry soil 6 inches in diameter and 12 inches long were placed with their feet standing one inch in water, under conditions where no evaporation could take place from their surfaces. Thus situated, no one of five samples was completely saturated at the end of 34 days. Before these soils appeared damp at the surface, the times given below were required: —

In clay loam,   time required to travel 11 inches, 6 days.
" reddish clay   "   "   "   "   11   "   22   "
"   "   "   "   "   "   "   11   "   18   "
" clay with sand "   "   "   "   11   "   6   "
" fine sand   "   "   "   "   11   "   2   "

During the first 24 days of this trial, the mean rate of capillary rise was .79 pounds per square foot daily, but when fully saturated, these soils were shown to lift the water at a rate exceeding 2.05 pounds per square foot and per day. In the case of the fine sand, the mean daily movement during the first 24 days was .69 pounds per square foot, but when fully saturated, it exceeded 2.37 pounds, a rate more than three times as fast.

When the soil grains are separated from one another, so as to develop an open, crumbly condition, then the rate of capillary rise of water through it is greatly reduced. Thus, plowing so thoroughly checks the loss of water from the soil beneath the stirred portion, that in one case seven very drying days failed to appreciably decrease the mean amount of water in the upper four feet of a field soil, while an immediately adjacent and entirely similar land, not plowed, lost, during the same time, the full equivalent of 1.75 inches of rain, or more than 9.13 pounds per square foot. So, too, in the case of the fine sand referred to above, while it was losing water at the surface at the rate of 1.38 pounds per day as a mean of ten days' trial, when two inches of the surface was removed and then laid directly back again, but in the loose, unfirmed condition, this treatment had the effect of reducing the loss of water at the surface, during the ten days immediately following, to a little less than .5 pounds per square foot daily, an amount, it will be seen, so small that the other rate exceeded it by more than 2.7 times.

The rate, too, at which surface tension or capillarity is able to move water through the soil is materially influenced by the presence of various substances dissolved in the soil water. The writer has found that when .08 per cent of potassium nitrate was added to distilled water, which was being lifted by capillarity through columns of rather coarse sand 18 inches long, the rate at which the water passed up and away by evaporation exceeded that with the pure distilled water by 22.84 per cent. The presence of lime water, of common salt, and of sulphate of lime or land plaster decreased the rate at which the water was lifted and evaporated from the surface, while potassium carbonate did not appreciably affect the rate of movement. A saturated solution of land plaster decreased the rise 27.36 per cent, while solutions of .08 per cent of common salt and potassium carbonate decreased the flow by 12.82 per cent and .66 per cent respectively.

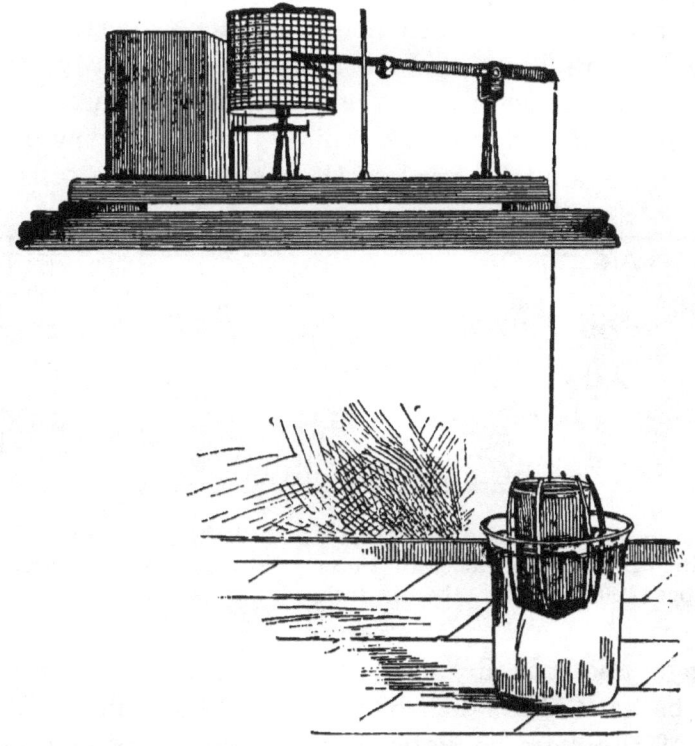

Fig. 24. — Showing self-recording apparatus for registering fluctuations of water in wells.

MOVEMENTS DUE TO SOIL AIR.

We have now to consider the movements of soil water due to expansions and contractions of the soil air. When a self-recording instrument similar to the one represented in Fig. 24 is placed so that its float is resting in the water of a well or spring, or, if it is properly placed, in the water discharging from a system of tile drains, it will be found that the water levels in all of these cases are subject to numerous and sometimes very complex movements. So, too, this instrument, suitably placed upon small lakes, and even rivers, shows their surfaces subject to oscillations of longer and shorter periods, many of which are associated with changes in air pressure. When the barometer falls, the water in wells rises, springs and tile drains flow more rapidly; but when the barometer rises again, the level of the water in wells falls, and

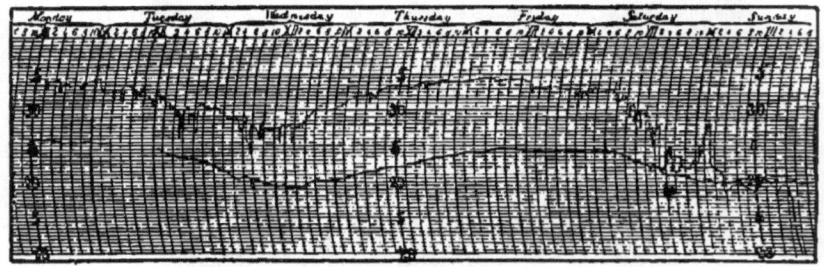

FIG. 25. — Showing changes in the rate of discharge of water from a tile drain coincident with those of barometric pressure. The upper line represents the tile drain.

springs and tile drains flow less rapidly. In Fig. 25 will be seen the synchronous changes of the barometer and of the flow of water from a tile drain as they occurred during one week.

To understand how changes in the barometer can affect the percolation of water into tile drains and into the natural waterways, it must be observed that there is always more or less air in the soil above the ground water, and that this air cannot easily and quickly escape, the soil offering resistance to its flow. Then, when the air pressure outside becomes less, this change is felt soonest at all natural openings to the ground water, such as wells, springs, and tile drains, and this lessening of the pressure at the outlets allows the water to escape more easily, and at the same time the air confined within the soil above the water table tends to expand and thus exert a downward pressure upon the water, tending to cause it to flow more rapidly and to percolate into tile drains and the natural outlets which lead to springs. But when the barometer rises again, when the air pressure becomes higher, then, as before, this pressure is felt first and strongest at places where the ground water is most exposed, and as a consequence some water is forced from wells back again into the soil which surrounds it, and the rate at which the water can emerge into tile drains and from the ground into springs is diminished for the time. These barometric changes have been observed to produce a change in the rate of flow of water from a tile drain amounting to 15 per cent, and in the case of a spring to 8 per cent.

Beginning in July and extending on through September, in Wisconsin, the surface of the ground water and the rate of percolation into tile drains are subject to diurnal oscillations in level and changes in the rate of flow which owe their origin to the daily expansion and contraction of air retained among the upper two to three feet of soil grains. The changes of pressure thus developed react upon the capillary water of the soil in the zone where the cavities are nearly or quite filled, forcing the water down and out into drainage channels when the air is expanding, but allowing such as has not been permanently lost to return again to its normal level when the pressure becomes less with the cooling and contraction

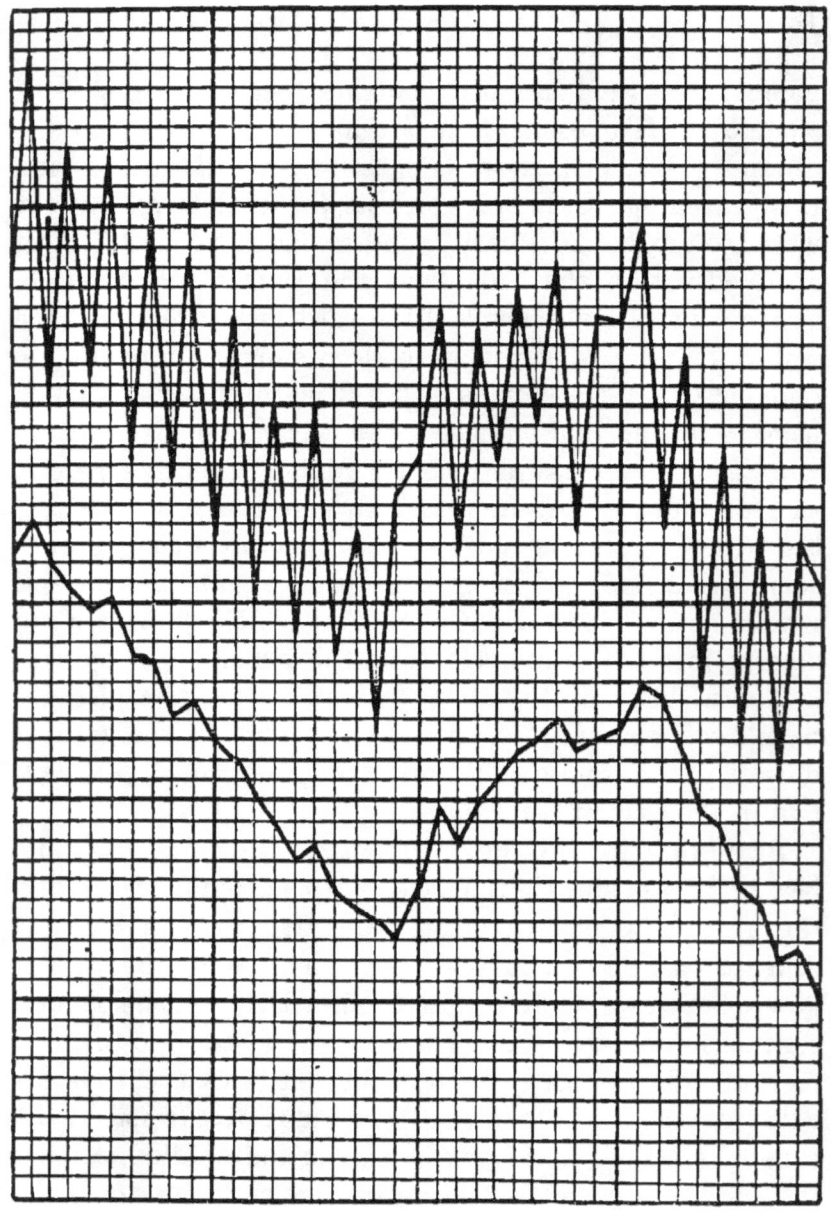

FIG. 26. — Showing diurnal changes of the ground water under conditions represented in Fig. 27. The lower curve represents changes in the inner well, and the upper one in the outer well.

of the soil air. In Fig. 26 are shown the diurnal changes here referred to as they occurred under the conditions represented in Fig. 27. It will be seen that the water

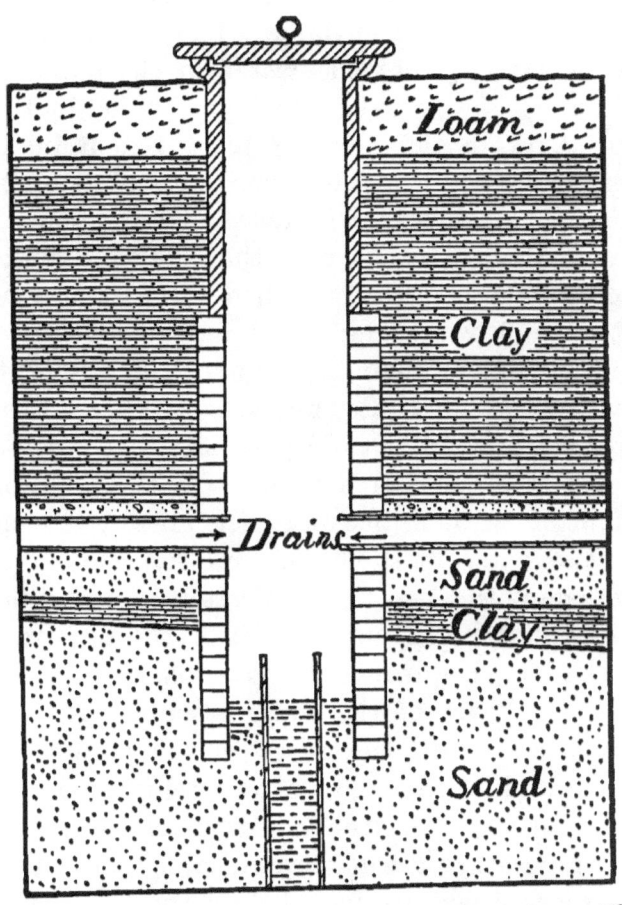

FIG. 27. — Showing the soil structure giving rise to the oscillations of ground water represented in Fig. 26.

in the well rose and fell with great regularity each day, and also with the slower changes of the barometer. The lower curve in this figure represents the changes in level as they occurred inside the 5-inch tile curbing in the centre of the main silt well, while the upper curve, where the diurnal changes are so pronounced, represents the changes in level as they occurred in the main well outside of the 5-inch tile curbing. The facts are, strange as they may seem, that the water in the outer well oscillated during the period of 20 days covered by the chart so as to stand, in the morning, from .1 to .3 inch above the level of the water in the inner one, and at night from .5 to 1.5 inches below that surface. To understand these movements, it must be observed that the well sunk in the centre of the larger one furnished the easiest escape for the water which was being lifted by the pressure of the surrounding higher water table under the adjacent hills, while vegetation in the vicinity of the well was using the water faster than it could percolate through the walls of the tile into the soil outside. On this account, and on account of the contraction of the air due to cooling of the soil, the water in the outer well fell each day, so as to stand below the level of the water in the inner one. But during the night, as the heat of the day was being slowly conducted downward, its expansion of the soil air forced out part of the water standing in the filled capillary spaces, until the water in the outer well rose from .1 to .3 inch above the water in the inner well. So far as the writer has observed, these diurnal changes due to temperature are not appreciable where the water table is more than 10 feet below the surface of the ground.

CHAPTER VI.

THE CONSERVATION OF SOIL MOISTURE.

To appreciate the need of conserving the soil moisture, it is important to know the relation which exists between the amount of rain received, and the possible loss of this rain in ways which do not contribute to crop production. Now, the waters which fall as rain upon a field may be lost to it by running off its surface, as is too often the case on hilly farms; it may be lost by percolating downward beyond the reach of root action; or it may be lost by surface evaporation from the ground itself. As a rule, it is only that water which passes through the plant which materially contributes to its growth. That which evaporates from the surface of the soil does indirectly help the plant by leaving the food it may have held in solution nearer the surface of the ground, where it is more likely to be taken up by the roots, and in assisting that soil life which develops fertility.

NEED OF CONSERVING SOIL MOISTURE.

We have no productive lands which, under field conditions, can retain as much as 30 pounds of water to the 100 pounds of soil, unless they lie close to or below the water table. In proof of this statement may be cited the water content of several field soils 32 hours after a rain of 3.19 inches which fell during 4 days.

Need of Conserving Soil Moisture.

These same fields had received 3.41 inches 20 days earlier, and during the intervening 20 days there had been a rainfall of 1.3 inches.

	First Foot. Per cent.	Pounds. Per cu. ft.	Second Foot. Per cent.
No. 1, Clay soil contained	33.91	26.79	26.60
No. 2, Clay loam "	28.88	26.57	26.45
No. 3, Clay loam "	28.75	26.45	27.06
No. 4, Sandy clay loam contained	26.48	25.95	25.46
No. 5, " " " "	24.69	24.76	22.81
No. 6, " " " "	24.61	24.12	21.44
No. 7, Sandy loam "	17.65	18.00	15.81
No. 8, " " "	17.65	18.00	14.59

In these cases the clay soil will weigh about 79 pounds per cubic foot, the clay loams 92 pounds, the sandy clay loams not more than 98 pounds, and the sandy loams about 102 pounds per cubic foot, so that the amounts of water retained in the surface foot in the several cases were as given in the table above.

It will be safe to say that those lands which can retain, in the upper five feet of soil, as much as 20 inches of water are very rare indeed. Then, were there no loss by surface evaporation nor by percolation downward, not more than 10 inches out of the supposed 20 inches of the stored water could be counted as available to a crop growing on the ground where large yields are expected; and it has been said that for lands to do their best, their water content should be steadily held up to from 40 to 50 per cent of saturation, and Hellriegel says 50 to 60 per cent.

In stating the rate of capillary movement of water in field soils, as measured by surface evaporation, we have seen that the daily losses may be as high as 1.3 to 1.6

pounds per square foot. On well-tilled, fallow ground, too, we have observed a mean loss of water amounting to .67 pounds per square foot, during a period of 64 summer days, which is equivalent to 8.24 inches. If we take the average daily loss from the soil at only .25 pounds per square foot, from May 1 to Sept. 1, this would be equivalent to 5.77 inches of rain, and deducting this from a mean rainfall of 12 inches for the same period, we should have left 6.23 inches for the use of a field crop, which, added to the 10 inches we have supposed might be withdrawn from that stored in the upper five feet of soil, gives a total for crop production of 16.23 inches.

Were it possible to have 16 inches of water to the acre for crop production, over and above losses by percolation and evaporation from the soil surface, large yields, so far as water contributes to them, would be certain,— yields amounting to from 3.5 to 7 tons of dry matter to the acre. In our experiments, where we have attempted to measure the water used in the production of a ton of dry matter, our smallest yield has been at the rate of 4 tons, and our largest at the rate of 17 tons, with an average for 22 cases of over 7 tons of dry matter per acre. In all of these trials, however, whether with oats, barley, corn, peas, clover, or potatoes, water was so added to the soil, from time to time, as to maintain for it a constant saturation of 40 to 50 per cent. Now, since ordinarily fertile lands, when well watered, will yield such large returns, and since it is seldom true that 16 inches of water are available for the use of crops under natural conditions, it follows that by whatever means we can diminish the loss of water from the soil, or render a larger per cent of it available to the crops growing upon it, their adoption may reasonably be expected to give larger yields. What, then, may be done to conserve soil moisture?

SAVING SOIL MOISTURE BY PLOWING.

Plowing land in the fall has a very appreciable influence on the per cent of water the surface three or four feet of such soil may contain the following spring, and the writer has observed a mean difference of 2.31 per cent more water in the upper three feet of immediately adjacent lands plowed late in the fall, as compared with that not plowed, the surface of neither having been disturbed until May 14. The larger quantity of water in the fall-plowed ground, in this case amounting to not less than 6 pounds to the square foot, was due partly to two causes; namely, the loose, open character of the overturned soil, causing it to act as a mulch during the fall, and again in the spring, after the snows had disappeared; and the more uneven surface, which tended to permit more of the melting snow and early spring rains to percolate into the soil.

Late fall plowing, leaving the surface uneven and the furrows in such a direction as to diminish washing, works in a decided manner, on rolling land, to hold the winter snows and rains where they fall, giving to such fields a more even distribution of soil water in the spring. And when it is observed that heavy lands, after a dry season, seldom become fully saturated with water during the winter and spring, the importance of fall plowing in such cases can be appreciated.

From the standpoint of large crops, which result from the best use of the soil moisture, there is no one thing more important for a farmer to strive for than the earliest possible stirring of the soil in the spring, after it has sufficiently dried so as not to suffer in texture from puddling. When the soil is wet, when its texture is close from the packing which has resulted from the winter snows and early spring rains, the loss of water is very rapid, as has been pointed out; it may be more than 20 tons daily per acre, and this loss may extend to depths exceeding four feet.

We must give here the details of an experiment aiming to measure the influence of early spring plowing on the loss of water from the soil, some of the results of which have already been referred to.

On April 28, a piece of corn ground was plowed and sowed to oats, and on the next day samples of soil were taken, in one-foot sections, down to a depth of four feet. Seven days later samples of soil were again taken, and below are given the percentages of water as they were found on the two dates.

	1st Foot. Per cent.	2d Foot. Per cent.	3d Foot. Per cent.	4th Foot. Per cent.
Water in soil April 29	20.13	22.61	20.51	17.72
Water in soil May 6	19.86	23.24	20.86	17.17
Difference,	−.27	+.63	+.35	−.55

It is thus seen that there had been little change in the water content of the soil during the week, the capillary movement upward nearly keeping pace with the loss by surface evaporation. Adjoining this piece and separated from it by a strip of grass only 10 feet wide, lay another piece of ground which was not plowed until May 6, and on this date, before plowing, samples of soil were taken from it in the manner already described, and below are given the amount of water which this unplowed land contained, and with these, for comparison, the amounts which were found in the piece which had been plowed seven days earlier.

	1st Foot. Lbs. Water.	2d Foot. Lbs. Water.	3d Foot. Lbs. Water.	4th Foot. Lbs. Water.
Land plowed	13.87	20.66	18.32	16.05
Land not plowed	10.58	17.98	17.28	13.94
Loss	3.29	2.68	1.05	2.11

It is thus seen that the unplowed ground had lost, during the seven days, in consequence of not having been plowed, not less than 9.13 pounds of water per square foot more than the plowed ground had, an amount equivalent to 1.75 inches of rain, and more than 198 tons of water per acre; and it should be said in this connection that there can be no reasonable doubt but that the actual loss was greater than this, because there must have been, during all this time, a movement of water upward from below the level at which the samples were taken into the four feet above, and whatever this movement may have been, it by so much diminished the observed losses.

There was another very serious result which followed this delay in plowing, for there were developed in the unplowed ground, as a consequence of the rapid loss of water from the soil, great numbers of very large and hard clods. So that, instead of having this piece of land in excellent tilth with plowing and once harrowing, it became necessary to go over the ground twice with a loaded harrow, twice with a disk harrow, and twice with a heavy roller, before it was brought into a condition of tilth even approximating what it might have had had it been plowed seven days earlier.

It follows from these observations that all lands, with but few exceptions, should be tilled at as early a date in the spring as their condition of moisture will permit, and not simply to save the all-important soil moisture and the formation of clods, but for other and very weighty reasons which will be stated in another place.

EARLY SEEDING AND CATCH CROPS.

The farmer who gets his crop to growing upon the ground at as early a date as the temperature of the air and of the soil will permit, is conserving soil moisture in still another way; for so soon as a crop gets possession of the land, so soon does it begin to use the water which is certainly running to waste upward, and possibly also downward; and more than this, the farmer who fails to give his crop possession of the land as soon as the season will permit, fails to get full advantage from the sheltering of the ground by the plants, and the diminished surface evaporation which results from the drying action of the crop. On all soils which do not have a strong tendency to run together after heavy rains, as a consequence of working, early surface tillage may often be adopted with great advantage even when the crops, like corn and potatoes, are not to be put in until late; a richer seed bed in better tilth, with more moisture and fewer weeds, being among the gains.

Where fields are naked through the winter, there is necessarily a large waste of moisture from the melting snows and earliest spring rains, which is often worse than a dead loss, because, so far as these waters drain away downward, they carry with them, and often beyond recovery, considerable amounts of nitrates and other valuable soluble and immediately available forms of plant food. Now, it is a matter worthy of serious consideration in studying this problem of conservation of soil moisture, whether it is not best in most cases of fall plowing to have some catch crop on the ground for the express purpose of utilizing the excess of moisture in the spring, which must be allowed to drain away or to evaporate before the ground is in condition to work. With the catch crop, the ground is dried earlier, and the moisture turned to account in preventing loss of fertility by drainage.

In the use of catch crops for this purpose, as in all cases of green-manuring, great judgment must always be exercised in order not to allow them to remain so long upon the ground as to so thoroughly exhaust the stock of soil moisture that the main crop is placed at a disadvantage on this account. In illustration of this danger, and as serving to enforce the great importance of keeping all fields free from weeds, an observation on the drying effect of clover will be instructive.

On May 13, the water content of soil in a field just planted to corn, and in a field of clover adjacent to it, was determined, the two sets of samples being taken within less than two rods of each other. The differences in the amount of water in the two cases were as given below.

	1 TO 6 INCHES. Per cent.	12 TO 18 INCHES. Per cent.	18 TO 24 INCHES. Per cent.
Corn ground	23.33	19.13	16.85
Clover ground	9.59	14.75	13.75
Difference	13.74	4.38	3.10

We have here a very forcible illustration of the danger of using catch crops, and of the evil consequences of allowing a field, when occupied with another crop, to become encumbered with weeds. In the case of the weeds, the soil is not only robbed of its moisture, but soluble plant food is also locked up in the tissues of the weed, in a form in which it is not immediately available. Had the clover field been plowed a week earlier than the samples were taken, it is quite likely that even then so much moisture had already been used that not enough remained to quickly decompose the material turned under, and at the same time meet the needs of the crop put upon the ground.

In plowing under coarse herbage, too, when the soil has a scanty supply of moisture, it is very difficult to establish such a capillary connection with the soil below as will allow a sufficient amount of water to rise into the overturned portion to start and properly feed a crop.

CULTIVATING AND HARROWING.

In the conservation of soil moisture by tillage, there is no way of developing a mulch more effective than that which is produced by a tool working in the manner of the plow, to completely remove a layer of soil and lay it down again, bottom up, in a loose, open condition. The harrow, when it scratches or cuts the surface of a field, without completely covering it with a layer of loose crumbs, hastens rather than retards the loss of moisture from the soil. So, too, when a tool is used which plows deep and wide grooves, leaving untouched and only partly covered ridges between them, it increases instead of lessening the rate of evaporation. That most excellent tool, the disk harrow, may be used so as either to hasten or check the loss of moisture from the soil, according as its disks are set at a very small or a large angle. In one trial, the writer increased the rate of evaporation from the surface of a 12-inch cylinder of soil from .75 pounds per square foot in 24 hours to 1.38 pounds in the same time by simply making vertical crosscuts in the surface with a sharp knife; but when a whole layer was removed in the fashion of the plow, and replaced in the loose condition, the rate of loss was then reduced to .5 pounds. When a strip of land is cultivated to a depth of 3 inches frequently during a whole season, while an adjacent strip is left smooth and unstirred, both pieces being fallow and kept free from weeds, the difference in the amount of moisture in the soil becomes very appreciable. In a test of this sort, where the amount of water in the soil was determined in one-foot sections to a depth of 6 feet, the mean daily loss per acre was at the rate of 14.48 tons on the cultivated ground, and 17.6 tons on the ground not cultivated, and these amounts are over and above what may have been brought up into the 6 feet by capillary action from below, and are the averages for 49 days. This represents a saving of water in that time equivalent to 1.7 inches of rainfall.

When stirring the soil to a depth of 3 inches is compared with that of 1 to 1.5 inches, the 3-inch mulch is decidedly more effective, so far as its conserving soil moisture is concerned. Through extended and repeated field trials, on different soils and in different seasons, the writer has found that invariably there is left at the end of the season a larger amount of water in the soil where it is stirred to a depth of 3 inches than when stirred to a depth less than this amount. The observed differences in a field of corn for one season are given below.

	1st Foot. Per cent.	2d Foot. Per cent.	3d Foot. Per cent.	4th Foot. Per cent.
Cultivated 3 inches deep	23.14	23.30	21.94	22.46
Cultivated 1 inch deep .	22.70	21.08	19.65	19.58
Difference .	.44	2.22	2.29	2.88

These differences at the end of the growing season, represent 167.4 tons more water per acre in favor of the deeper mulch.

SOIL MULCHES.

When the effectiveness of very thin soil mulches is measured, it is found that here, too, the lessening of the rate of evaporation may be very considerable. Thus it was found, by laboratory methods, that when the evaporation from an unstirred surface was at the rate of 6.24 tons per acre daily, from the same surface, when stirred to a depth of .5 inch and .75 inch, the loss was only 5.73 tons and 4.52 tons respectively. Again, when the surfaces were covered with a fine, dry clay loam to a depth of .5 inch and .75 inch, the daily loss was then at the rate of 6.33 tons per acre per day for the naked surface, but only 4.54 tons and 2.4 tons for the mulches, respectively.

There is thus left no room to doubt the efficacy of dry earth mulches as conservators of soil moisture, but it should be said that not all soils are equally effective in their power to diminish evaporation.

The writer has found that, taking the evaporation through a mulch of 2 inches of dry quartz sand, which passes a screen of 20 meshes, but is retained by one of 40 meshes to the inch, as 1, the rate of evaporation through the same depth of finely pulverized, air-dry clay loam was 3.5 times more rapid; the latter giving, in the still air of the laboratory, a loss of 3.474 inches in depth per 100 days, while the loss through the sand was .995 inch in the same time. This difference in the effectiveness of the two soils to act as mulches depends apparently upon the difference in their capillary power, the fine soil with its small pores, even when air-dry, lifting the water faster than the sand.

It is to be observed, further, that the mulching effect of a soil decreases with time, the capillary power becoming stronger as the per cent of water in the mulch increases. Experiments showed that the coarse sand referred to above, which had been allowing water to pass through it during 30 days at the mean rate of .027 inch, decreased the rate to .0099 inch when a fresh mulch was substituted. So, too, in the case of the fine clay the rate of evaporation decreased from a mean daily loss of .071 inch to .034 inch when the old mulch was replaced by the fresh. It follows, therefore, that to maintain the most effective mulches in the field requires frequent stirring, even though a rain may not have occurred to pack the soil.

TRANSLOCATION OF SOIL MOISTURE.

When a surface soil has its water content reduced, so that the upper 6 to 12 inches is beginning to be dry, the rate of capillary rise of water through it is decreased and it begins to assume the properties of a mulch. But when this condition has been reached, if a rain increases the thickness of the water film on the soil grains without causing percolation, the capillary flow may be so strengthened that the surface foot draws upon the deeper soil moisture at a more rapid rate than before, causing a translocation of the lower soil moisture, the deeper soil becoming measurably drier soon after such a rain than it was before it, while the surface foot is found to contain more water than has fallen upon it.

The following experiment may be cited in proof of this important principle. At 5.30 P.M. samples of soil were taken on a piece of fallow ground in one-foot sections to a depth of four feet. Water was then applied to this surface at the rate of 1.33 pounds per square foot. Samples of soil were also taken adjacent to this wetted area to serve as a control experiment, and 19 hours later corresponding sets of samples were again taken with the results stated below.

POUNDS OF WATER PER CUBIC FEET OF WET AREA.

	1st Foot.	2d Foot.	3d Foot.	4th Foot.
Before wetting	11.78	15.79	14.73	14.03
After wetting	14.06	17.52	15.58	15.40
Gain	2.28	1.73	.85	1.43

POUNDS OF WATER PER CUBIC FOOT OF AREA NOT WET.

	1st Foot.	2d Foot.	3d Foot.	4th Foot.
First samples	12.38	17.05	14.92	14.48
Second samples	12.75	17.72	15.40	14.17
Gain	.37	.67	.48	−.31

Now, it will be seen from these results that the water content of the soil increased on both areas, but at the rate of 6.23 pounds per square foot on the wet area, and 1.21 pounds on the area not wet. The 19 hours which intervened between the taking of the two sets of samples was a period of very little evaporation, most of it being in the night, and the following morning was cloudy and very damp, and the result was that capillarity gave to the area not wet 1.21 pounds more water per square foot than it lost by evaporation. But the wet area had gained 6.23 pounds, and yet only 1.33 pounds had been added to the surface, making the increase by capillarity

$$6.23 - 1.33 = 4.90 \text{ lbs.};$$

and if we subtract from this the amount which the control area gained, we shall have 3.69 pounds as the water of translocation due to the wetting of the surface.

This is not an isolated observation; for the experiment was twice repeated, and the fact had been several times observed in the field when taking samples of soil just before and again immediately after rains, these observations leading to making the experiments.

Now, when it is recognized that rains which do not cause percolation tend to strengthen the flow of the deeper soil water toward the surface, while they have at the same time greatly diminished the power of this soil to act as a mulch, it should be very evident that, at critical times like these, no time should be lost in developing a new mulch which shall retain in the upper soil, not simply the rain which has fallen, but the moisture it has caused to be brought up from below; for if this is not done, the ground as a whole will have become drier in a few days than it would have been had no rain fallen.

It will often happen in farm practice, after a field of corn or of potatoes has been laid by in perfect condition, so far as being free from weeds and in possessing a good mulch are concerned, that a rain may come, making it advisable to cultivate the field once more in order to restore the mulch, and to retain the water which translocation has brought up within the reach of root action.

It will be evident also that when watering transplanted trees or other plants, during a dry time, care should be taken to add water enough to produce percolation, and then to protect the wet surface with some sort of mulch; for unless this is done, the chances are that more harm than good will result. It is evident, too, from what has been said regarding mulches, that transplanted trees should be early protected by an ample mulch and by keeping the ground above and near the roots entirely free from grass and weeds, whose power to withdraw water from a dry soil far exceeds that of the pruned and mutilated roots of transplanted trees.

There is another effect resulting from the translocation of soil moisture, as influenced by cultivation and the use of mulches in other ways, which is very important in the saving and utilization of the water deep in the ground. It is in the upper 18 inches of soil that the nitrates are developed through the processes of nitrification, and it is from here, too, that plants must obtain them. To the upper soil also other fertilizers are applied, and from it these must be removed, and it can only be done when water enough is present to allow the processes of diffusion to be carried forward.

Now, keeping the upper 18 inches of soil sufficiently moist enables the process of translocation to bring into use a larger amount of the water deep in the ground;

that is, to lift it into the zone of soil near the surface, where, through a more plentiful supply of air, carbon dioxide and the presence of micro-organisms, it is far more serviceable in preparing and giving to the plant the food it needs than it could be if absorbed by the roots at a lower level in the ground.

That a sufficiently deep mulch, prepared by cultivation, does act in this manner is proved by the observed distribution of water in the soil when cultivated deep and shallow and when cultivated and left unstirred. In both of these classes of cases the writer has repeatedly observed at a certain stage in the drying of field soils that, while the surface two or three feet of the better mulched soil will contain more water than the corresponding layer of the unmulched or poorly mulched soil, the third and fourth or the fifth and sixth feet may be wetter beneath, where the surface has been least protected.

On three different fields of corn, with as many different kinds of soil, where the ground was cultivated 3 inches and 1.5 inches deep respectively, the writer found on July 16, 1894, the following distribution of soil moisture.

	1st Foot. Per cent.	2d Foot. Per cent.	3d Foot. Per cent.	4th Foot. Per cent.
No. 1, cultivated 3 inches	11.30	15.57	10.54	11.37
No. 1, " 1.5 "	9.92	15.43	11.56	13.99
Difference,	+1.38	+.14	−1.02	−1.62
No. 2, cultivated 3 "	13.96	22.74	23.39	19.47
No. 2, " 1.5 "	12.98	20.44	24.02	21.34
Difference,	+.98	+2.30	−.63	−1.87
No. 3, cultivated 3 "	11.65	17.47	16.44	13.03
No. 3, " 1.5 "	10.65	16.85	17.81	13.32
Difference,	+1.00	+.62	−1.37	−.29

Differences similar to these were obtained in cornfields in 1893 and also on fallow ground, where the rates of evaporation on cultivated and uncultivated soil were compared, and that these differences are effective in crop production is borne out by the fact that in fifteen trials out of twenty the yield of corn was larger on land cultivated 3 inches deep than it was on land cultivated only 1.5 inches deep. The facts appear to be that wherever we can succeed in holding the per cent of water relatively high in the surface two feet, a larger per cent of the water in the next four feet at least becomes available and is actually brought nearer the surface, where it has the highest value in crop production; and then, if what has been said is true, it follows that less water and less fertility are lost by percolation.

EFFECT OF ROLLING ON SOIL MOISTURE.

It used to be generally believed by farmers, and perhaps the majority of them still have the impression, that firming the surface of the ground, as with the roller or with the wheels of the press drill, increases the water content of the soil; and that this belief should have been so generally entertained is what would be expected from the very evident fact that firming the surface of the ground does, for the time, increase the amount of water in the compacted portion.

When, however, the changes in the water content of the surface four feet of soil which follow the use of a heavy roller are studied, it is found that we have here another case of the translocation of soil moisture; a case where, by destroying the many large non-capillary pores in the soil and bringing its grains more closely

Effect of Rolling on Soil Moisture.

together, its water-lifting power is increased and to such an extent that often within 24 hours after rolling, the upper one or two feet of the firmed ground have come to contain more moisture than similar and immediately adjacent land does at the same level, while the lower two feet have become drier. Water has been lifted from the lower into the upper soil.

In the table below will be seen the difference in the water content of the soils which have been rolled, and immediately adjacent ones not so treated. These results are averages derived from 147 pairs of samples.

Surface	36 to 54 inches,	unrolled	19.43	per	cent.
"	36 to 54 "	rolled	18.72	"	"
		difference	−.71	"	"
"	24 "	unrolled	19.85	"	"
"	24 "	rolled	19.49	"	"
		difference	−.36	"	"
"	2 to 18 "	unrolled	15.64	"	"
"	2 to 18 "	rolled	15.85	"	"
		difference	+.21	"	"

It is here seen that when samples are taken to a depth exceeding two feet, the rolled ground as a whole is drier than that not rolled, and that this difference is greater when the samples are taken to depths of from three to four or more feet. The data presented also show that the surface 2 to 18 inches of loose ground recently firmed contains more water than that which has not been so treated.

While firming loose soil tends, therefore, at first, to increase the moisture in the surface soil at the expense of the deeper layers, the whole ground soon comes to contain less, on account of an increased rate of evaporation.

It is not simply because the water is brought to the surface by firming that a more rapid loss of water takes place after rolling, but making the ground very smooth increases the wind velocity close to the surface, and to such an extent that it may exceed that on the unrolled ground by more than 70 per cent. It is plain, therefore, that whenever it becomes desirable to firm the surface for the purpose of increasing the amount of water in it, a light harrow should follow the roller in order to restore a thin mulch which shall retain the water brought up by the firming; for unless it is done, only a temporary advantage is derived from it.

It sometimes occurs, after sowing spring grain, that heavy rains develop a crust, which diminishes the porosity of the soil beyond what is most desirable. When this has taken place, and if the surface is uneven, owing to harrow ridges and numerous lumps, then, by using the roller when the ground is dry, the crust and the lumps may be crushed and a partial mulch formed, while at the same time the porosity of the surface will be to some extent restored.

DEEP TILLAGE AND LEVEL CULTURE.

In the conservation of rains in regions where they are scanty and where surface evaporation is very rapid, it may sometimes be desirable to work the soil unusually deep in order to insure a deeper percolation of water into the ground and a slower return of it to the surface. Deep working, through the formation of numerous non-capillary spaces, increases the rate of percolation and at the same time decreases the capillary return to the surface; so that in climates where so little water falls that only a small amount ever drains away, methods of tillage may be desirable which in humid regions would not be prudent. In regions of large rainfall, at times when fields are not bearing crops, the great danger lies in the loss of fertility through excessive percolation; so that in such countries and at such times a rapid evaporation of soil moisture is to be encouraged.

Ample root room is a very necessary condition for the largest utilization of soil moisture. When, for any reason, the roots of plants are forced to develop close to or near the surface of the ground, as is the case when a field is insufficiently underdrained, then, as the plants approach the fruiting stage and begin to use water very rapidly, the zone of soil in which the roots lie is so rapidly depleted of its moisture that capillarity is unable to prevent it from becoming dry, and the result is that a large amount of moisture near at hand becomes unavailable, when, if the water table in the early part of the season had occupied a lower level, the roots would have developed more deeply, and they could have come into near contact with a larger volume of soil and of water. Then by taking water at a relatively slower rate and from a greater depth, a strong capillary flow upward is longer maintained and from lower levels in the ground, so that a much larger amount of the soil moisture becomes available.

Then, too, those methods of tillage which leave the surface of the field nearly flat rather than thrown up into ridges and hills are less wasteful of soil moisture. To hill potatoes or corn to a height of 6 inches when the rows are 3 feet apart may increase the surface exposed to the sun and evaporation more than 5 per cent, and if ridged to a height of 9 inches more than 9 per cent. Under these conditions the water must rise to a greater height under the rows before reaching the surface roots, while midway between them and where the ground is least shaded, the unmulched surface lies nearest the water supply. These being the conditions, ridge culture must be more wasteful of soil water than level tillage, whence it becomes evident that naturally dry soils everywhere, and most soils in dry climates, should, wherever practicable, be given flat cultivation.

On stiff, heavy soils in wet climates and during wet seasons it may become desirable to practice ridge culture with potatoes and some of the root crops, but not so much to increase the rate of evaporation from the soil as to provide a soil bed in which it will be less difficult for fleshy tubers and roots which form beneath the surface to expand. A clay soil, after having been thoroughly wet and then allowed to dry, shrinks into so firm a mass that it is very difficult for potatoes to expand in it; and under these conditions they tend to form at or above the surface. Hilling in such cases is very beneficial, although more wasteful of soil moisture.

DESTRUCTIVE EFFECT OF WINDS.

In arid or semiarid countries and in districts where the soil is light and leachy, but especially where there are large tracts of land whose incoherent soils suffer from the drifting action of winds, it is important that the velocity of the winds near the ground should be reduced to the minimum. We have in Wisconsin extensive areas of light lands which are now being developed for purposes of potato culture; but while these lands are giving fair yields of potatoes of good quality, they are in many places suffering great injury from the destructive effects of winds. On these lands, wherever broad, open fields lie unprotected by windbreaks of any sort, the clearing west and northwest winds after storms often sweep entirely away crops of grain after they are 4 inches high, uncovering the roots by the removal of from 1 to 3 inches of the surface soil. It has been observed, however, that such slight barriers as fences and even fields of grass afford a marked protection against drifting for several hundred feet to the leeward of them.

In the case of groves, hedgerows, and fields of grass, their protection results partly through their tendency to render the air which passes across them cooler and more moist, and partly by diminishing the surface velocity of the wind. The writer has observed that when the rate of evaporation at 20, 40, and 60 feet to the leeward of a grove of black oak 15 to 20 feet high was 11.5 cc., 11.6 cc., and 11.9 cc., respectively, from a wet surface of 27 square inches, it was 14.5, 14.2, and 14.7 at 280, 300, and 320 feet distance, or 24 per cent greater at the three outer stations than at the nearer ones. So, too, a scanty hedgerow produced observed differences in the rate of evaporation, as follows, during an interval of one hour: —

At 20 feet from the hedgerow the evaporation was 10.3 cc.
At 150   "   "   "   "   "   "   "   12.5 cc.
At 300   "   "   "   "   "   "   "   13.4 cc.

Here the drying effect of the wind at 300 feet was 30 per cent greater than at 20 feet, and 7 per cent greater than at 150 feet from the hedge.

When the air came across a clover field 780 feet wide, the observed rates of evaporation were, —

At 20 feet from clover, 9.3 cc.
At 150   "   "   "   12.1 cc.
At 300   "   "   "   13.0 cc.

or 40 per cent greater at 300 feet away than at 20 feet, and 7.4 per cent greater than at 150 feet.

The protective influence of grass lands and the disadvantage of very broad fields of these light soils was further shown by the increasingly poorer stand of young clover as the eastern margin of these fields was approached, even on fields where the drifting had been inappreciable. Below are given the number of clover plants per equal areas on three different farms, as the distance to the eastward of grass fields increased:—

No. 1, at 50 feet, 574 plants; at 200 feet, 390 plants; at 400 feet, 231 plants.

No. 2, at 100 feet, 249 plants; at 200 feet, 277 plants; at 400 feet, 193 plants; at 600 feet, 189 plants; at 800 feet, 138 plants; at 1000 feet, 48 plants.

No. 3, at 50 feet, 1130 plants; at 400 feet, 600 plants; at 700 feet, 543 plants.

In these cases the difference in stand appears to have resulted from an increasing drying action of the wind. On the majority of fields the destructive effects of the winds were very evident to the eye, and augmented as the distance from the windbreaks increased.

It appears from these observations, and from the protection against drifting which is afforded by grass fields, hedgerows, and groves, that a system of rotation should be followed on such lands, which avoids broad, continuous fields. The fields should be laid out in narrow lands and alternate ones kept in clover and grass. Windbreaks of suitable trees must also have a beneficial effect when maintained in narrow belts along line fences and railroads and perhaps wagon roads, in places.

CHAPTER VII.

THE DISTRIBUTION OF ROOTS IN THE SOIL.

WHEN we look at the habit of growth among higher plants, it is to be noted that, excepting a few very consolidated forms of vegetation like the cacti, whose natural habitat is in arid or semiarid regions, plants spread out in the air and in the soil two very broad surfaces joined by a relatively narrow and rigid stem. The roots, branching in the ground, dividing and subdividing until numberless root hairs have threaded themselves through the capillary spaces among the soil grains, are there to gather water, nitrogen, and ash ingredients for growth. So large is the quantity of water demanded by plants, so small is the amount of water within a small area of soil, and so slow is the method by which the roots obtain it, that nothing short of an enormous root surface could do the work, and how great this surface really is may be partially appreciated from the photo-engraving of the roots of corn, clover, oats, and barley shown in Fig. 28. These roots were obtained by growing the plants in a deep cylinder holding between 500 and 600 pounds of earth and then, at maturity, carefully washing the soil away with a fine stream of water.

THE NEED AND GREAT EXTENT OF ROOT SURFACE.

That we may the more clearly appreciate the great need there is for the vast extent of root surface spread out by agricultural crops, and how important it is that there shall be a deep, well-drained, and well-tilled soil in which they may expand, let me give the measured amounts of water used by four stalks of corn and withdrawn by their roots from the soil between July 29 and August 11. Two stalks of maize were growing in each of

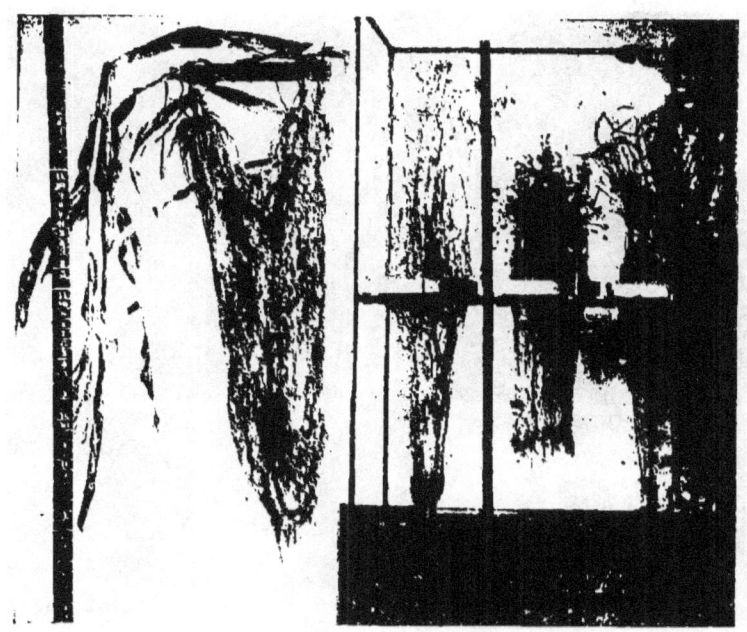

FIG. 28. — Showing the total root of four stalks of maize, and of oats, clover, and barley.

two cylinders filled with soil, having a depth of 42 and a diameter of 18 inches. These four stalks of corn, as they were coming into tassel and their ears were forming, used during 13 days 150.6 pounds of water, or at the mean daily rate of 2.896 pounds for each stalk. Had an acre of ground been planted to corn in rows 3 feet 8 inches each way and four stalks in a hill, then, with an average consumption of water at the observed rate given above, there would have been withdrawn from that acre an amount of water during those 13 days equal to 244 tons or 2.42 acre-inches; and when it is observed that this must be withdrawn from a soil so dry that no amount of pressure could express from it a drop of water, it is not

FIG. 29. — Showing the natural distribution of corn roots in a field soil under natural conditions.

strange that a mass of roots like those shown in Fig. 28 should be required to do the work with sufficient rapidity.

Referring now to Fig. 29, it will be seen how completely the whole soil of the field is threaded with roots; for in both cases two hills of corn, standing opposite each other in adjacent rows, are shown and the roots meet and pass one another between the hills, and in the younger stage these had already exceeded a depth of two feet, while in the second case, taken just as the corn was coming into tassel, the roots had descended until at this time the whole upper three feet of the field soil appeared to be so fully occupied with corn roots that not a cube of earth one inch on a side existed in the three feet of depth which was not penetrated by more than one fibre of thread-like size. In many portions of the soil the roots were much closer than this, and the minute root hairs which branch out from the thread-like fibres referred to, and which constitute the chief absorbing surfaces of the roots, are not included in making this statement of root occupancy.

At the distance apart of planting in the field from which these roots were taken, there were in the surface three feet $40\frac{1}{3}$ cubic feet of soil available for each four stalks, so by multiplying the 1728 cubic inches in one cubic foot by $40\frac{1}{3}$, the number of cubic feet of soil occupied, we get a total of 69,696 cubic inches. If, then, each cubic inch of this soil contains not less than one linear inch of thread-like root, their aggregate length could not be less than one-twelfth of 69,696, or 5808 feet, which is just 1.1 miles.

Let the reader keep in mind that the corn roots here under consideration grew in the field under perfectly natural conditions, and that the cage of wire shown in the engraving was simply slipped over the block of soil which contained the roots there shown, after the corn had reached the stage of maturity represented in the figure. It should also be understood that the four stalks of corn which drank the 150.6 pounds of water in 13 days, did it at the stage of growth represented by the oldest plants in Fig. 29; and, further, that these stalks were only good average plants, such as would make a yield of 4.5 tons of dry matter per acre.

Is this not a sublime expression of one of nature's methods of utilizing all her opportunities and making the most out of whatever is at hand? The rocks had fallen into decay, the sun warmed the bed of incoherent fragments, the winds had brought the rains, capillarity had retained a portion of them on the surfaces of the soil grains, the forces of solution had charged these waters with plant food, but roots were needed by which plants could utilize these resources. As man, standing on the bank of the stream and looking for means to better his condition, was led to put his wheel into the water and withdraw some of the energy running by unused, so have plants, during the long ages of fitting and refitting, learned to build into the capillary and percolation streams which flow past the grains of soil such wonderful systems of absorbing surfaces as we have seen that greatest of American food plants, the maize, to possess, and of which America's most revered poet wrote, in 1855: —

>Day by day did Hiawatha
>Go to wait and watch beside it;
>Kept the dark mold soft above it,
>Kept it clean from weeds and insects,
>
>*   *   *   *   *
>
>Till at length a small green feather
>From the earth shot slowly upward,
>Then another and another,
>And before the summer ended
>Stood the maize in all its beauty,
>With its shining robes above it,
>And its long soft yellow tresses;

> And in rapture Hiawatha
> Cried aloud, "It is Mondamin!
> Yes, the friend of man, Mondamin!"
> Then he called to old Nokomis
> And Iagoo, the great boaster,
> Showed them where the maize was growing,
> Told them of his wondrous vision,
> Of his wrestling and his triumph,
> Of this new gift to the nations,
> Which should be their food forever.
> And still later, when the Autumn
> Changed the long green leaves to yellow,
> And the soft and juicy kernels
> Grew like wampum hard and yellow,
> Then the ripened ears he gathered,
> Stripped the withered husks from off them,
> As he once had stripped the wrestler,
> Gave the first feast of Mondamin,
> And made known unto the people
> This new gift of the Great Spirit.

Looking at Fig. 30, which shows the roots of winter wheat, barley, oats, blue grass, timothy, clover, and another leguminous plant, *Lathyrus sylvestris*, it will be observed that with these plants, all except the blue grass send their roots into the fourth foot of soil, and in these cases, as with the corn, there is no portion of soil in the upper three or four feet which a root can penetrate that is not threaded by roots much closer together than one in every cubic inch. Not all of the roots belonging to the tops shown in the figure just referred to were retained, but simply those which grew vertically downward within the area of a circle 12 inches in diameter.

How great may be the mass of roots associated with the tops of clover, oats, and barley, is clearly shown in

Need and Great Extent of Root Surface. 213

Fig. 28, where the total root system of these plants was recovered from the soil in which they grew. It is very probable that, had the cylinders been deeper in which these plants were grown, their roots would have attained a greater length, and this would certainly have been the

Fig. 30. — Showing the vertical distribution, under field conditions, of the roots of blue grass, timothy, Lathyrus sylvestris, clover, winter wheat, barley, and oats.

case with the clover, whose roots matted thickly at the bottom of the barrel in which they grew.

Looking at the roots of winter wheat in Fig. 30, it will be seen that a mass of other roots have been entangled with them at a depth of 2 to 3 feet below the surface. These are from a second-growth black oak standing in a pasture 33 feet distant from the place where the sample of wheat roots were taken. In Fig. 31 may be seen the roots from a smaller tree of the same species as they were drawn from a sandy soil with the aid of a stump puller. In digging in a field adjacent to an apple orchard, the writer has found roots of the apple as large in diameter as a slender lead pencil at a distance of 45 feet from the trunk of the tree to which it was attached.

Fig. 31. — Showing the roots of second-growth black oak as drawn from a sandy soil.

Such facts as these illustrate in a forcible manner how deeply and broadly the roots of plants are sent foraging through the soil, and how much soil it is needful for them to come in contact with in order to procure the food and water they need.

Schubart, a German farmer, Storer tells us, made measurements of various roots which he washed out

Need and Great Extent of Root Surface. 215

from the soil of fields where the crops were growing. He found the roots of wheat, sowed in September, extending to a depth of 7 Rhenish feet in a subsoil composed of sandy loam; and Lawes observed the roots of lucern penetrating to a depth of 9 feet below the surface.

Referring again to the roots of maize which we recovered from the field at different stages of maturity, it may be stated here, as having an important bearing on the depth of cultivation of this crop, that as the corn advances towards maturity a portion of the roots which are thrown off at higher levels on the stem develop more nearly horizontally and come closer and closer to the surface, making it more difficult to cultivate deep late in the season than when the corn is small.

At the time when the corn had a height of about 18 inches, July 9, the roots in the centre between the rows 3.5 feet apart were nearly 8 inches below the surface, rising in a festoon to near the surface of the ground at the hill on either side. At this time, too, most of the roots were confined to the upper 18 inches of soil. When the corn had attained a height of 2.5 to 3 feet, the surface leaders, midway between the rows, had risen to within 6 inches of the top of the ground and now occupied the upper 2 feet of soil. Just as the corn was coming into tassel, the upper leaders were then scarcely 5 inches below, while at maturity they were less than 4 inches from the surface.

It is evident from these facts that corn may safely be cultivated more deeply early in the season than when it is more mature.

Each of the trunk roots or leaders of the corn plant sends out on opposite sides, much as the stalk does its leaves, slender rootlets from 2 to 6 inches long, and those on the surface leaders rise directly upward and nearly reach the top of the ground in the latter part of the season.

Those soils which are sandy and loamy in character rather than clayey, and whose grains have little tendency to draw together into block-like masses of varying sizes, as do the stiff clays, allow a much more symmetrical development of the roots in them, and as a consequence of this there is less trespassing of one root upon another; there is not so much soil far removed from the root hairs, and hence the soil water is more easily, rapidly, and completely removed for the purposes of the plant, and finally in these soils the roots find less difficulty in wedging the soil grains apart as they need more room in their growth.

In the clayey soils, too, as they shrink and crack, drawing themselves together in small cube-like blocks, there is a tendency, during dry seasons, to tear or break off many of the smaller rootlets, and thus deprive the plant of its means of water supply when water is most needed, and when, if the roots were left intact, it might be had.

There seems to be operative in the plant a power or directive influence, which leads to the most rapid growth of roots in those directions in the soil where the most bountiful supplies of food and the best conditions for growth exist.

When cultivated fields lie along one side of a grove or row of trees, these trees develop the strongest roots beneath the tilled fields, in just the same manner as the branches grow most rapidly in the direction of the most sunshine and the largest amount of room. Whether the lines of most rapid growth are determined solely by, and are the result of the most complete feeding, — that is, whether those roots which lie where the most food materials in the form of water and ash ingredients can diffuse into them, are in precisely the condition which causes the assimilated material from the stem to be brought to the roots to make a more rapid growth possible, — need not be discussed here. It is true, however, that a plant does not usually waste its energies in developing roots in a direction where no benefit could be derived from it.

CHAPTER VIII.

SOIL TEMPERATURE.

There is no physiological fact more evident than the extreme importance of maintaining the right temperature surroundings for living forms of all kinds. In our own case a deviation of the mean temperature of the body a few degrees either above or below 98° F. results in very serious consequences, if long continued. So imperative is the right bodily temperature in many animals, and so narrow is the admissible limit of variation, that it is necessary to have in the body compensating devices which are automatically under the control of the nervous system. When we are in health if, for any reason, the mean temperature of the body is becoming too high, the sweat glands are set at work and the surface of the body covered with perspiration, the evaporation of which withdraws so much heat from the body that the temperature is brought down to the normal point. Then, should the temperature tend to fall too low, the activity of the skin is diminished, and at the same time a larger amount of oxygen is taken into the body to unite with the food eaten, for the definite object of producing heat more rapidly.

In the vegetable world the extremes of temperature which are destructive to life are usually much farther apart than they are among animals; that is, the temperature may be allowed to fall much lower and to rise much higher with no other serious consequences than the partial or complete cessation of physiological processes, but usually here there is both an upper and a lower limit, beyond which life is destroyed.

It is not difficult to understand why these temperature relations should be so essential, and from the practical point of view it is very important that they should be. Many of the essential changes which take place in the body of a plant or of an animal, and which constitute the life processes, are chemical in their character, as is that of the burning of wood. But when we desire wood to burn, in ordinary air, when we wish oxygen to unite with it chemically, we have first to apply heat to it and raise its temperature up to a certain point before burning will begin. When the chemical action has been once started, then there is usually heat enough produced to maintain the action so long as there is a plentiful supply of oxygen and of wood. Then, when water is thrown upon a fire to extinguish it, it does so chiefly by withdrawing heat from the flames so rapidly in the evaporation of the water that the temperature is lowered to such an extent as to stop the chemical action. The steam formed by the evaporation also occupies so much space as to greatly dilute the oxygen, and in this way tends to make the burning slower or to stop it altogether.

To light the match, we rub it vigorously upon some rough surface, but this is only to raise its temperature up to the point at which oxygen will begin to unite with it; to warm the prepared end in any other way would produce the same result, and the sole reason for using the preparation at the end of the match is because chemical action can be started in it at a temperature so low that it can easily be produced by friction.

Now the chemical changes in the animal body and in

the tissues of plants, which result in the various phenomena of growth and manifestations of power, require that the materials which react one upon another should first be raised to a certain temperature, should have a certain rate of molecular swing, before it is possible for the changes to take place. Just as water evaporates the faster the more heat is communicated to it whereby its molecules, by absorbing that heat, are thrown outside of the liquid mass and become a gas; and just as the fuel in the stove must be warmed up to a certain temperature before the force of cohesion is enough weakened to allow the oxygen of the air to unite with it; so must the materials within the seed, in the roots, stems, and leaves of plants have their temperature raised to a certain point before those chemical and physical changes which constitute the phenomena of growth can take place.

IMPORTANCE OF SOIL WARMTH.

Now the lowest soil temperature, according to Ebermayer, at which the processes of growth are started in most cultivated crops is from 45° to 48° F., but the maximum results are attained only after the soil has reached a temperature of 68° to 70°. Let us compare these temperatures with those observed in the soil at different depths, for the six months, April to September inclusive. Dr. Frear reports the soil temperatures at State College, Pennsylvania, from which we derive the mean of five years, as given in the table below:—

Depth.	April.	May.	June.	July.	Aug.	Sept.
3 inches	43.74	55.13	67.29	70.16	68.70	61.32
6 "	43.08	54.72	66.34	69.75	68.49	61.70
12 "	42.69	53.83	65.03	68.89	68.66	62.73
24 "	41.43	51.45	61.90	66.42	67.41	63.59

These temperatures, it will be observed, for the month of April, are below the minimum for growth given above, while the August temperatures only barely reach 68° F.

Ebermayer, at Munich, found the temperatures in a loess-like loam as follows, for a mean of four years: —

Depth.	April.	May.	June.	July.	Aug.	Sept.
5.9 inches	44.65	56.79	61.11	67.26	64.09	58.21
11.8 "	44.31	57.51	60.06	66.16	63.61	57.88
23.7 "	44.40	53.58	59.11	63.12	63.55	58.82
35.4 "	43.56	51.24	57.83	62.92	62.26	58.51

These temperatures, it will be observed, are higher in April and May, but lower the balance of the season, than they are in Pennsylvania.

Now if it is true that the vital processes of plant life can only go forward rapidly and normally when the surrounding temperatures are right, it is evident that this factor in the culture of plants must be of even greater importance than it is in the successful management of animals; for the latter have it within their power to raise or lower their own bodily temperature as the surrounding conditions may demand, while this power is not to any notable degree possessed by plants.

Let us then consider some of the ways in which too low or too high soil temperatures may work to the disadvantage of a crop growing upon it. It may be stated at the outset that in temperate climates, where the soils are supplied with the needed amount of moisture, there is little likelihood that the temperature will become too high for the majority of our cultivated crops. The danger lies in the direction that the soil will be too cold.

If we desire to dissolve almost any substance quickly in water, or indeed in almost any acid, we can greatly hasten the solution by using heat; and from what has been said regarding the method of solution, this should be expected. But one of the chief functions of soil water is to take into solution, from the soil constituents, substances which, for the most part, dissolve, under the most favorable conditions, very slowly. Now, raising the temperature of the soil grains, weakens the attractive force which holds the plant food locked in the solid state, while at the same time it makes the diffusion of the dissolved materials, away from the seat of action, more rapid; and unless the dissolved materials are removed as soon as formed, they, by their reaction against the face of the soil grains, tend to prevent any further action from taking place.

It is not only important that there should be a rapid diffusion of the dissolved food away from the places where it is being formed, through the action of soil water, but it is even more important that this process should be carried forward in the soil air, in order that a sufficient amount of fresh oxygen and nitrogen should enter the soil to replace that which is being used by roots, seeds, micro-organisms, and chemical processes needful to a fertile soil; and the higher the soil temperature is, the greater are the absolute velocities of the molecules of air, and the faster will they reach the place in the soil where they are needed, and the sooner will the rapidly forming carbon dioxide escape into the air, leaving room for more to form. A high soil temperature, then, is conducive to a more rapid and thorough soil aëration or ventilation, a process of extreme importance, as we shall see in another place.

It is not enough that a rapid solution of plant food shall take place in the soil, but in order that growth may be vigorous, it is necessary that the dissolved food should be transported quickly to the places in the plant where it is to be used. Now the strength of osmotic pressure, and the rate of its action, increases as the temperature of the medium in which it is taking place is higher. We must keep in mind that in the process of osmosis, as in the diffusion of gases or in the evaporation of water, the molecules go from place to place by virtue of the rate at which they are thrown by the heat imparted to them, and hence, if the soil temperature is held high, the molecules of water are hurled into the root hairs and on into the other tissues faster than they can be if the surrounding temperature, the supply of power, is low; but when the molecules are once within the plant cells, then the higher speed they possess in virtue of their higher temperature is just the condition which develops the strong osmotic pressure required to force the sap onward toward and into the leaves, no matter how high the stem which bears them may be.

Sachs has shown, for example, in the case of both the tobacco plant and the pumpkin, that they wilted even at night and with an abundance of moisture in the soil, so soon as the soil temperature fell much below 55° F. Under this temperature the power which moved the water from the roots to the leaves was too feeble to compensate for the slow evaporation which takes place at night.

Then, again, when seeds germinate in the soil, work must be done, and in no small degree, at the expense of the heat absorbed by the soil. Haberlandt found, for example, that the germination of wheat, rye, oats, and flax goes forward most rapidly at 77° to 87.8° F., and that corn and pumpkins germinate best at 92° to 101° F. He found that when corn would germinate in three days at a temperature of 65.3° F., it required eleven days when the temperature of the soil was as low as 51° F. He found, further, that when oats would germinate in two days impelled by a temperature of 65.3° F., it required seven days to do the same work when the temperature was as low as 41° F.

These are forcible illustrations of the need of a warm soil, and should be a sufficient spur to every thoughtful farmer to do whatever he can to meet the conditions here made evident. That certain seeds require a higher soil temperature for their germination than others do, must be understood as meaning that the necessary transformations do not take place so easily. It must not be understood, however, that when a seed is placed in a cold soil, the food stored in it for the development of the young plant cannot undergo a change. Quite the contrary; there are organisms in the soil which are able to do their work at low temperatures, and if a seed, under these conditions, comes into their presence, it absorbs moisture and, being unable to grow, becomes a prey to these lower forms and decays.

In speaking of the sources of nitrogen for higher plants, it was stated that the nitrates formed their chief supply. But in studying the conditions under which the nitric ferment works most vigorously, it has been learned that the germs cease to develop nitric acid from humus when the temperature falls below 41° F.; that its action is only appreciable at 54° F., while it becomes most vigorous at 98° F., but that at 113° F. its activity drops back again to what it was at 59° F. Here, again, is another and very urgent need for the right soil temperature.

What, then, are the conditions which influence soil

temperatures? And, since a soil is more often too cold than too warm, what can be done to raise its temperature?

CONDITIONS INFLUENCING SOIL WARMTH.

There is no one cause so effective in holding the temperature of a soil down as the water which it contains, and which may be evaporating from its surface. This is because more work must be done to raise the temperature of one pound of water through one degree than of almost any other substance. Thus, while 100 units of heat must be used to raise 100 pounds of water from 32° to 33° F., only 19.09 units, according to R. Ulrich, are required to warm the same weight of dry sand, and 22.43 units an equal weight of pure clay, through the same range of temperature. To raise the temperature of 100 pounds of dry humus through 1° F., it is necessary to give to it 44.31 heat units, while 100 pounds of carbonate of lime require 20.82 units.

From these figures it is evident that when the sun imparts equivalent amounts of heat to equal weights of sand, clay, humus, and water, the sand will be the warmest, while the water will be the coldest. To make the differences definite, suppose the water has its temperature raised 10° F., then the same amount of heat entering an equal weight of humus will make it 22.6° warmer, clay 44.58° and sand 52.38° warmer. But while the temperatures of these soils would stand in the relation of the figures here given when they are dry, it is not true that under field conditions such large differences of temperature would be observed, because there are other factors which modify the effect of differences of specific heat, whose influence alone we have thus far considered.

Since the weights of dry soils per cubic foot are not the same, it is evident that the heaviest soil, were other conditions alike, would be the coolest, because while the same amount of sunshine can fall upon a square foot of sandy soil as falls upon an equal area of clay soil, the cubic foot of sand weighs 110 pounds, while that of clay may only weigh 75 pounds; hence, a surface foot of sandy soil, instead of being 7.8° warmer, would be 8.6° colder than the clay soil. But the fact that the water-holding power of the clay soil is greater than that of the sandy soil, works again in the opposite direction, on account of the large amount of heat needed to warm the water, to make the clay colder than the sand; so that, if we assume the clay and sand saturated, the same number of heat units per surface foot of each soil would give the sand a temperature 3° F. warmer than the clay.

The chief cause, however, which makes a wet or undrained clay soil colder than a well-drained or sandy soil is the large amount of heat which is used up in evaporating the excess of water from the surface. While 100 heat units will raise the temperature of one pound of water through 100° F., it is necessary to use 966.6 heat units to evaporate one pound of water from the soil; but this, if withdrawn directly from the cubic foot of saturated clay, would lower its temperature about 10.3° F. It must be evident therefore, that to allow the surplus water to drain away from a field rapidly, rather than to hold it there until it has time to evaporate, must greatly favor the warming of the soil.

The writer has observed the following differences of temperature in the surface inch of a well-drained sandy loam and an undrained black marsh soil, both of them naked and level.

Date.	Time.	Condition of Weather.	Temperature of Air.	Temperature of Drained Soil.	Temperature of Undrained Soil.	Difference.
Apr. 24	3.30 to 4.00 p.m.	Cloudy, with brisk east wind.	60.5° F.	66.5°	54.00°	12.50°
Apr. 25	3.00 to 3.30 p.m.	Cloudy, with brisk east wind.	64.0° F.	70.0°	58.00°	12.00°
Apr. 26	1.30 to 2.00 p.m.	Cloudy, rain all the forenoon.	45.0° F.	50.0°	44.00°	6.00°
Apr. 27	1.30 to 2.00 p.m.	Cloudy and sunshine, wind S.W. brisk.	53.0° F.	55.0°	50.75°	4.25°
Apr. 28	7.00 to 8.30 a.m.	Cloudy and sunshine, wind N.W. brisk.	45.0° F.	47.0°	44.50°	2.50°

It should be noted in connection with this table that the differences in temperature in favor of the drier soil have occurred when the amounts of evaporation would be relatively small, and hence when the differences would be below rather than above the average.

Comparing the temperatures of a well-drained clay soil with that of a sandy loam, also well drained, when each was less than half saturated, the following differences were observed on August 6 : —

	1st Foot.	2d Foot.	3d Foot.
Sandy loam	76.5° F.	74.7° F.	72.1° F.
Clay loam	69.5°	69.3°	67.0°
Difference,	7.0°	5.4°	5.1°

Now, from what has been said in regard to the advantages of a warm soil, it is plain why a sandy loam, when well fertilized, is for so many purposes superior to the heavy clay soils.

The slope of the land surface, and the direction of this slope with reference to the points of the compass, often have a marked effect upon the temperature even when the soils are identical. Thus, on the south shore of Lake Superior the writer found the temperature of a stiff red clay on a level table and on a south exposure sloping about 18° to have the following values on July 31 : —

	1st Foot.	2d Foot.	3d Foot.
Red clay, south slope	70.3° F.	68.1° F.	66.4° F.
" " level	67.2°	65.4°	63.6°
Difference,	3.1°	2.7°	2.8°

In a level, sandy, alluvial soil close by the above and at the same time, the temperatures were 71.2° F. 70.1°, and 67.6° for the first, second, and third feet respectively. Here, again, the influence of the character of the soil on its temperature is well marked, but the differences are smaller than in the former case, as the tabulation shows. In these two cases both soils were more than half saturated with water : —

	1st Foot.	2d Foot.	3d Foot.
Alluvial sand, level	71.2° F.	70.1° F.	67.6° F.
Red clay, level	67.2°	65.4°	63.6°
	4.0°	4.7°	4.0°

Wollny, in his study of soil temperatures as affected by slope and direction of exposure to the sky, found, through observations on small artificial hills with inclinations of 15° and 30° to the horizon, that the south side has an average temperature of 1.5° F. when the slope was 15°, and 3.1° F. with a slope of 30°, warmer than the north side; but, comparing the east with the west slope, he found less than .2° F. difference. Comparing the east and west slopes with the south, he found, for 15° inclina-

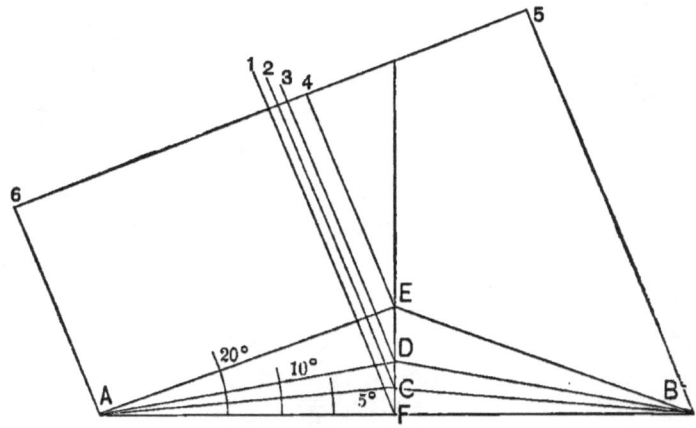

Fig. 32.—Showing how the slope of the surface influences the amount of heat received per unit area.

tion, the east .71° F. and the west .56° F. colder than the south side; but when the 30° slopes were compared, the differences were 1.31° F. for the east and 1.44° F. for the west colder than the south slope.

Just why a south slope should be warmer than a north slope will be readily seen from an inspection of Fig. 32. Suppose *A 6 5 B* represents a prism of sunshine falling upon the hill *AEB*, where *AE* is the south slope and *EB* the north. Here it will be seen that, on account of the sun not being directly above the hill, the south slope gets as much more sunshine than the north slope does as the line *46* is longer than the line *45* which, for the angle of 20°, is about one-third more per each square foot of surface.

The color of a soil, too, has not a little influence in determining its temperature, the darker soils, if other conditions are the same, being the warmer. All are familiar with the fact that a black garment is much warmer in bright sunshine than a white one. This is because the black surface absorbs the ether waves as they come from the sun in much larger proportions than the white garments do. The leaves which drift upon the white snow, and the dirt also, as all have observed, cause the snow to melt much more rapidly than where the snow is clean and white. So the sandy soils which are light in color do not become as warm as they might if their light color and glistening surface did not cause much of the sunlight to be turned directly back without doing any work upon them.

A rough, uneven, lumpy, or ridged surface has a tendency to make the soil colder than the smoother, more nearly level surfaces do, and there are several causes which co-operate to produce these results. If the surface is ridged east and west, or if it is covered with lumps, the south sides of these inequalities receive more heat than the north sides do, causing them to become relatively very warm, and in this condition they lose much more heat by radiating it away and by the air coming in contact with and being warmed by them.

The effect of rolling the land on the temperature of the soil is often very marked, its general tendency being to make it warmer during bright clear weather, but in cloudy and cold weather it has the opposite effect, rolled land tending to cool more rapidly. The writer has found through extended studies, under field conditions on soils of various kinds, that a rolled field may have a temperature, at 1.5 inches below the surface, as much as 10° F. above entirely similar soil not so treated, and at a depth of 3 inches, a difference of 6.5° has been observed. The differences in temperature observed on one date are shown in Fig. 33. Here, it will be seen, the air temperature

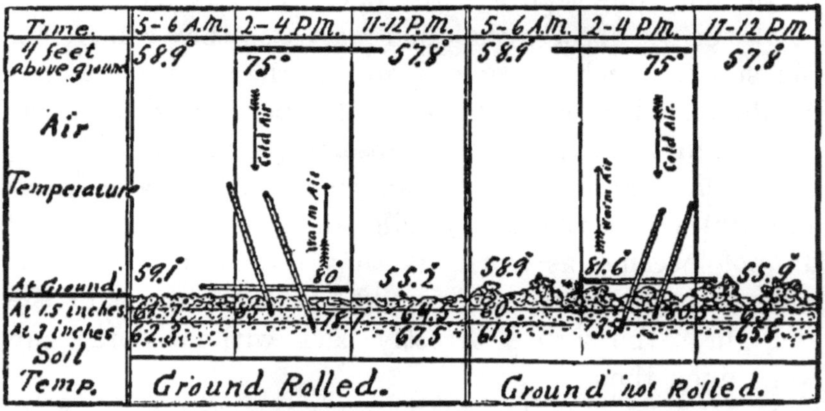

Time.	5-6 A.M.	2-4 P.M.	11-12 P.M.	5-6 A.M.	2-4 P.M.	11-12 P.M.
4 feet above ground	58.9°	75°	57.8°	58.9°	75°	57.8°
Air Temperature						
At Ground.	59.1	80	55.2	58.9	81.6	55.9
At 1.5 inches	60.7		64.5	80		60.0
At 3 inches Soil Temp.	62.3		67.5	61.5	73	65.5
	Ground Rolled.			Ground not Rolled.		

FIG. 33. — Showing observed differences of temperature on rolled and on unrolled ground.

over the unrolled ground is warmer, except early in the morning, than it is over the rolled surface, showing that the dry lumps and uneven surface are imparting their heat to the air more rapidly, and, since both surfaces are receiving the same amounts of heat from the sun, it is plain that if the air is warmed more over the unrolled ground the soil must be warmed less. The fact that the surface soil is less firmly packed on the unrolled ground also causes the heat to be conducted downward less rapidly, and thus tends to make the deeper soil cooler. During the night and cold, cloudy weather, when little heat is being received from the sun, the loose soil on the unrolled land acts as a blanket and tends at such times to make the rolled land the coolest at the surface, as shown in the figure.

The observed temperatures between 1 P.M. and 4 P.M., on eight Wisconsin farms, are given below: —

Soil.	Air Temperature.	Soil Temperatures.			
		At 1.5 Inches.		At 3 Inches.	
		Rolled.	Unrolled.	Rolled.	Unrolled.
	F.	F.	F.	F.	F.
Sandy	56.75°	65.55°	63.50°	61.92°	59.45°
Clay loam	51.17°	56.55°	53.60°	56.48°	53.55°
" "	59.00°	70.21°	64.40°	64.85°	59.90°
Sandy	69.62°	73.40°	72.05°	70.79°	69.26°
Clay	72.64°	79.77°	78.20°	73.02°	71.82°
Sandy clay	73.30°	76.83°	71.90°	72.58°	69.09°
Yellow clay	60.80°	61.34°	58.82°	57.38°	55.58°
Clay loam	79.70°	89.89°	86.12°	81.49°	76.47°
Mean	65.37	71.69	68.57	67.31	64.39
Difference	. . .	3.12	. . .	2.92	. . .

These results were all obtained in the spring, on land sowed to grain, and show an average difference of 3.1° F. and 2.9° F. higher temperature on the rolled ground.

The character and depth of cultivation also has an appreciable effect upon the soil temperature, stirring it deeply, tending to make the deeper soil cooler than where the depth of cultivation is less. This tendency is due to the fact that the loose soil of greater depth is a poorer conductor of heat, and tends to shut out the heat from the sun. The temperatures where corn ground was cultivated 3 inches deep and 1.5 inches deep were found by the writer to be .82° F. warmer in the surface foot, .59° F. in the second foot and .36° warmer in the third foot on land cultivated 1.5 inches deep than they were on adjacent and similar soils cultivated to a depth of 3 inches. These differences were determined by a self-recording maximum and minimum thermometer, which had a bulb one foot long, enabling it to record the mean temperature of the foot of soil in which it was placed, and the differences given above are the means of measurements in eleven fields in different parts of Wisconsin.

It is also true that the fermentations of organic matter which go on in the soil under the influence of various microscopic organisms, produce no inconsiderable amount of heat, and for this reason, among others, a field heavily fertilized with farmyard or green manures enjoys a higher soil temperature because of the treatment.

There is perhaps no method of soil warming at considerable depths in the ground so effective as that of warm, percolating rains. In the spring of the year the soil is usually thoroughly saturated with the water from the winter rains and melting snows, and under these conditions, when the warm spring rains come they usually percolate deeply into the ground, shoving the colder soil water in front of them beyond the depth of the root zone. When we recall that each pound of rain at 60° F. carries 10 heat units into the ground, which can be applied to raising the temperature of soil colder than 50° F. up to that point, and that each such heat unit can raise the temperature of a pound of sand 5.24°, the great advantage of warm April showers in bringing an early spring can be appreciated. But of course, when cold rains come to the fields, the opposite effect must result.

It is evident therefore, from what has been said regarding the temperature of soils, that not only is the right degree of warmth very important, but the farmer has it within his power at times to make the soil warmer or colder. It will be remembered that the vital processes of plant life only begin after the temperature has risen above 45° to 48° F., while the mean April temperatures of soils cited on a preceding page fall below these figures, and since the earlier we can bring the soil up to the growing temperature the larger will be the conservation of both soil moisture and other plant food, it is evident that if we can, without undue expense, hasten the warming of the soil to a practical extent, it is important that it should be done.

MEANS OF CONTROLLING SOIL WARMTH.

Now the thorough preparation of the seed bed which good farmers so much insist upon is justified in a large measure by the warming effect which judicious, thorough tillage has. In the first place, the development of a mulch over the ground lessens the loss of water from the surface by evaporation, and if by this means we have saved from loss into the air 10 pounds of water per square foot, we have avoided by this saving the withdrawal of 10 times 966.6 heat units from the same area; so that early thorough tillage not only saves moisture, but it at the same time permits the seed bed to become quickly both dry and warm enough to allow nitrification to begin and to hurry forward. It is very important, too, that this should be done; for the melting snows and spring rains have carried down beyond the reach of young plants most of the soluble nitrates the soil may have possessed the fall before. Early thorough tillage, then, develops to an important extent needed plant food by warming and aërating the soil, while at the same time it hastens the germination and gets the plant into condition to take advantage of the food being prepared.

It is not so important early in the spring that the ground should be warmed deeply, but better only so far as to provide ample available plant food for the start, and room enough for the first roots, and this is what early tillage does; the loose, open soil conducts neither the cold ground water up to become still colder by evaporation, nor the absorbed sunshine down where it is not yet needed. In effect this method converts the whole field into a hotbed or cold frame, and commences irrigation by saving water at the start. It does even more; for while it develops early a little soluble plant food, it holds in abeyance these processes in the deeper soil, which, by starting too early, would cause a needless loss by developing nitrates which may be leached away before the crop has its roots ready to use them.

We have pointed out how and to what extent rolling warms the soil. Should the farmer use his roller to warm his soil? Yes, at times; but he must do so with good judgment. From what has been said, it follows that rolling may warm the ground too deeply, and at the same time waste soil moisture. Evidently the practice to follow, so far as the problem here under consideration is concerned, is to wait after using the tool only so long as to permit enough moisture to reach the surface, and enough warmth to pass downward to meet the immediate needs, and then the harrow should follow without a moment's delay.

A roller, to do its work properly, must have a considerable weight, and the larger the diameter of the roller, the heavier it should be. The plank can seldom be used as a substitute for the roller, when firming the ground is what is demanded. This will be readily understood, when it is recalled that the much lighter weight the plank must have is spread out over many times the extent of surface covered by the roller, and hence its pressure, per square foot, must be proportionately less. The roller presses, with its whole weight concentrated, along a narrow line, and the narrower the smaller the diameter of the roller. But the smaller the diameter of the roller, the harder it must draw in proportion to its weight.

When a mellow, open seed bed has been prepared, and its temperature has been raised to the proper point, should a rain fall upon it, that water will tend to pass through its wide pores quickly to the deeper soil, and without leaching it as badly as would be the case were the soil more compact; so that in the early season when there is an over-abundance of moisture, it is best, for warmth, for aëration, and to lessen loss of fertility by percolation, to have a mellow seed bed.

It seems to the writer probable that there may be times when we should till the surface for the express purpose of keeping the deeper soil cool even late in the season, when it has become desirable that it should have a good degree of warmth. Reference has already been made to the diurnal changes in the ground water, due to diurnal oscillations of soil temperature. But under certain conditions these oscillations result in a loss of soil moisture through increased drainage, and with this an increased loss of plant food.

In Fig. 34 are shown automatic records of the diurnal changes in the level of the ground water in a tile-drained field, and of the rate of flow of water from the system of tiles which was carrying the percolating waters away. The highest portions of these curves mark the

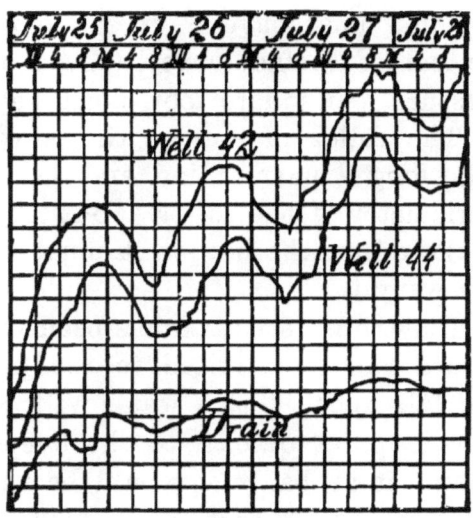

FIG. 34.—Showing the diurnal oscillations of the water table and of the rate of flow of water from a tile drain due to diurnal temperature changes.

time when the water was flowing slowest, and when the water in the wells stood at the lowest level, while the lower parts of the curves mark the time when the water was flowing fastest from the tile drains, when percolation was most rapid. Referring now to Fig. 35, which shows the changes in soil temperature 18 inches below the surface, and of the air temperature above, it will be seen that the warmest time in the soil is a little after midnight, while it is coldest a little after noon. The most rapid flow of water from the drain, and the highest level of the water in the wells, occur about 7 o'clock in the morning, hence some hours later than the highest soil temperature; and since it is the expansion of the soil air, due to the rising temperature, which produces this effect, a certain amount of lagging should be expected.

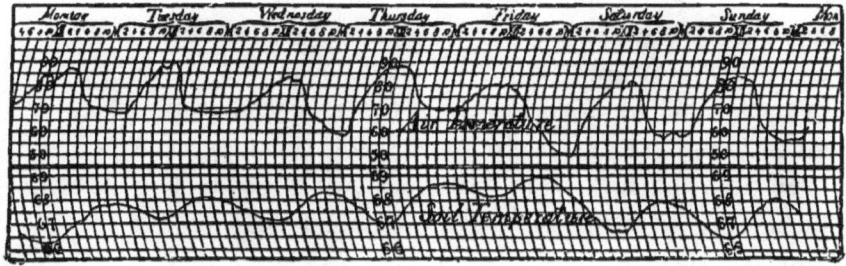

FIG. 35.—Showing diurnal changes in the soil temperature at 18 inches below the surface and of air temperatures one foot above the surface, as given by thermograph.

Now in such localities as these, and possibly also where the ground water is not as near the surface, the developing over the field of a loose, non-conducting mulch of soil, produced by cultivation, must make the diurnal changes of temperature less, and hence the percolation of the soil water also, saving the water for the crop.

CHAPTER IX.

THE RELATION OF AIR TO SOIL.

The presence of an ample amount of air in the soil is as indispensable to the life of upland plants as is that of water, and whatever method of tillage is adopted, it should not hinder soil breathing to an injurious extent.

It has been abundantly demonstrated that when free oxygen is completely excluded from seeds placed under otherwise good conditions for germination, growth will not take place; if after seeds have commenced to sprout the oxygen supply is cut off, they cease to develop. It is true that the germination of seeds will take place in an atmosphere very poor in oxygen, but after the percentage amount has been reduced below $\frac{1}{32}$ of the normal quantity, growth is much retarded and sickly plants usually are the result.

NEEDS OF SOIL VENTILATION.

Practical experience teaches that when a soil bearing other than swamp vegetation is flooded, or even if it is kept long with its pores filled with water, the plants soon sicken and die, and this, too, when they are in full leaf and abundantly supplied with food and warmth. The difficulty is the lack of root breathing; the plants are drowned, and as effectually as an animal would be under water, because enough free oxygen cannot reach them.

It is true, to be sure, that on the floating gardens of the Chinese and the Mexicans, crops are matured under conditions where the roots of the plants must be immersed in water during their whole period of growth, and this fact appears to be a contradiction of the statements just made. In these cases, great rafts of basket-work are covered with soil, and floated upon a lake or stream, so that here the roots of whatever crop the floating islands may have produced must have occupied saturated soil or the water itself below the raft. So, too, in the cases of water culture, where plants have been grown without soil, in water holding in solution the needed nitrogen and ash ingredients, there is an apparent contradiction. In these cases, however, the water is free from the soil, where, by absorption and diffusion, oxygen from the air can enter it even more readily than it is able to do in a good soil, for there a large part of the space is occupied by impenetrable soil grains; and more than this, wind and convection currents are powerful adjuncts in bringing fresh oxygen to the submerged roots of the plants just as they are in bringing to the fish and other animals the oxygen they need.

In the compact, water-filled soil, however, all current motion is prohibited, and the roots of plants can only secure the free oxygen which the water may have brought with it when it fell as rain, but as this amount is small, it soon becomes exhausted. On the other hand, if the field is underlaid by a deep, porous subsoil, into which the rains may quickly penetrate, or if the field is underdrained, then, as the water runs away, leaving empty spaces behind, air must be drawn in to take the place, and needed oxygen is thus supplied.

We have seen in another place, that the germs which develop nitric acid in the soil find oxygen indispensable to their life, and so important is a large supply of it in the soil that in olden times, when saltpetre farming was practiced to procure potassium nitrate for use in the manufacture of gunpowder, great pains were taken to thoroughly aërate the soil in which the nitrate was being developed, and as one of the chief objects of tillage is the production of nitrates in the soil, we may profitably recall the old method of raising saltpetre.

In the growing of nitrate of potash in Europe, a mixture of soil, manure, and leached ashes or marl was formed into beds, sometimes made more open and porous by the use of gratings or racks. Great care was taken to keep the beds warm and of the proper degree of moisture, while from time to time they were shoveled over or otherwise stirred for the purpose of introducing air and warmth. Then when a new field or bed was to be started, care was taken to bring to it soil from an old nitre bed, or, in the language of that time, adding "mother of petre," a term so appropriate in the light of our knowledge of to-day that nothing more significant could have been used. It is thus seen that the old nitre farming found its best results, its largest yields of saltpetre, in a rich soil kept well moistened, well stirred, and thoroughly warmed, all of them conditions which we now recognize as very important for any crop, and we cannot doubt that they are so because they favor the rapid development of that important plant food, nitric acid.

We have just been speaking of the need of oxygen in the soil to carry forward the process of nitrification.

It must now be said that oxygen is also needed to prevent the destruction of the nitrates after they have once been formed. From Warington's review of our acquaintance with the phenomena of denitrification, we learn that the destruction of nitrates with the setting free of nitrogen in waters containing sewage was observed by Dr. Angus Smith in 1867, and in 1868 Schlösing showed that nitrogen gas, or some of its lower oxides, are set free in several putrefactive and fermentive processes. He also found that when a moist soil, rich in humus, was kept in an atmosphere of free nitrogen, from which all oxygen had been excluded, all nitrates which they may have contained quickly disappeared. He found, further, that the same denitrifying process took place when a limited quantity of ordinary air was present. It was even true with the soils rich in organic matter that they often gave off more free nitrogen than was represented by the nitrates present in them; that is, other organic nitrogen was broken down and set free, this process consuming the oxygen derived from the decomposed nitric acid.

With these facts before us, it is plain that we are in danger of having the soil depleted of its needful nitrates, not only by excessive leaching, but also through the destruction of the organic matter from which these are evolved, if the land is allowed to remain too long with insufficient ventilation, as the result of poor drainage.

Warington conducted experiments in the Rothamsted laboratory on an artificially water-logged soil to which he had applied sodium nitrate at the rate of 519 pounds per acre, and found that, in less than three weeks, all but 21 per cent of the nitrate had disappeared, its oxygen having been used in the carrying on of life processes for which free oxygen would have been used had it been present in the soil in sufficient quantity. We must see to it, then, that our soils are sufficiently ventilated to insure the production of nitrates on the one hand, and to prevent their destruction on the other.

Now that we know how free nitrogen from the air is fixed in the tubercles on the roots of leguminous plants, it is evident that here is another reason why air must be admitted to the soil, and not simply to the surface layers which are tilled, but deeply into the root zone two and three or more feet. It will be seen that oxygen must be admitted to the soil in order that the nitrogen of decaying organic matter may be converted into forms available to higher plants, while free atmospheric nitrogen is demanded to hold up the stock of organic nitrogen by making good the losses in the drainage waters and by the processes of denitrification.

Those fermenting processes which result in the return of carbon as carbon dioxide to the air after it has answered the purposes of living tissues, require, many of them, oxygen from the air; hence, in order that the vast root systems which grow and die in the soil may not accumulate unduly, air in sufficient quantity must penetrate deeply. We have sufficient evidence of the advantage of good ventilation in the strong heating of the well-aërated heaps of horse manure, when contrasted with the much slower fermentation which takes place in close cow dung, free from litter.

There are many purely chemical reactions essential to soil production and soil fertility which demand a certain measure of oxygen and carbon dioxide for their continuance, so that here is another need for a fair measure of openness in the soil. And then if oxygen must be admitted to the soil for the setting free of nitrogen and carbon dioxide, it is equally necessary that these gases shall be allowed to escape with sufficient freedom, so that they shall not exclude the atmospheric air, or so dilute it as to render it ineffective.

We have now in mind the chief needs for a sufficiently free passage of atmospheric air in and out of the soil. How, then, is soil ventilation accomplished? How is it hindered, and what may we do to control it?

NATURAL PROCESSES OF SOIL VENTILATION.

The most general and constant mode of interchange of gases in the soil, and in the air above, is that of diffusion, whose motive power is the sunshine absorbed by the surface layers. As the molecular motion of the soil grains increases during the day, as the temperature rises, a part is transmitted to the gases contained in its pores, and more of these molecules are driven out than enter it; but when the soil temperature falls, then the reverse condition takes place, so that we have in the upper layers of the soil an incessant temperature breathing. But if the surface of the soil did not warm and cool by turns, there would still be an interchange of soil and atmospheric gases so long as their compositions were not identical; for, owing to the never-ending to-and-fro motion of the air molecules, wherever the oxygen is being consumed in the soil, there is produced at that place an oxygen vacuum into which other molecules are sure to find their way, unless the soil pores are in some way blocked to them. On the other hand, if carbon dioxide is being produced in the soil at any place, then an excess of pressure of this gas results, which pushes some of the molecules out into the open air.

We have already seen how the temperature effect in the soil is felt even in the deeper ground water and to such an extent as to force it out into drainage channels. So, too, if for any reason carbon dioxide is more rapidly set free than it can escape, the increased pressure which results will affect the ground water in the same manner as the temperature changes do. In like manner a more rapid consumption of oxygen or nitrogen in the soil than is compensated for by inflow from above tends to develop a negative pressure on the soil water, which would permit capillarity to return to the soil spaces water which had been forced into the beginnings of drainage channels or passages.

The aëration of the deeper soil is favored by the fact that, owing to the lagging of the diurnal changes of temperature to which reference has been made, the second foot may be growing warmer at the same time the surface foot is becoming colder, and *vice versa*. The second foot, for example, being warmest at the same time the surface foot is approaching its lowest temperature, more air from the second foot is forced into the surface foot by diffusion than would occur were there no temperature lagging. Then in the daytime, when the surface foot is becoming warmest, while at the same time the second foot is growing colder, the air then diffuses more rapidly downward into the deeper soil.

Every change which takes place in the barometer or in the atmospheric pressure above a field has a tendency to cause some air to pass either into or out from the soil. These alterations of pressure in the soil air are so marked that the movement of the ground water is very sensibly affected by them, as will be seen by referring to Fig. 36, which shows the changes in the rate of flow of water from a spring as they were modified by variations in the atmospheric pressure. In Fig. 37, too, may be seen the changes of level of water in a well during 24 hours, most of which were coincident with fluctuations in atmospheric

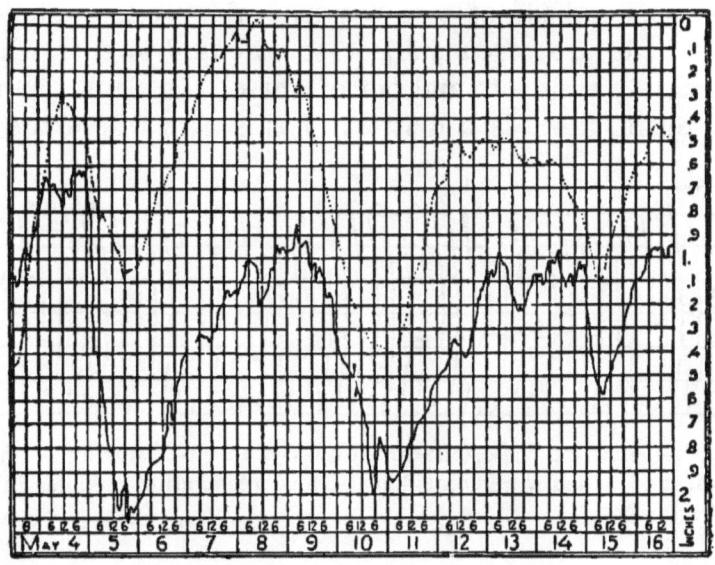

FIG. 36. — Showing fluctuations in the rate of flow of water from a spring at Whitewater, Wisconsin, May 4 to May 16, and the barograph record at Madison, Wisconsin, for the same period. Both reduced to natural scale. Heavy curve represents the spring.

pressure. But every change in atmospheric pressure which is large enough to affect the level of water in wells must also be large enough to cause some air to enter or leave the surface layers of soil.

When the wind is blowing strongly across the surface of the ground, but with greatly varying velocities, as is usually the case, there is a tendency to alternately suck air out from the soil pores, and again to allow it to return, thus producing an irregular but sometimes very strong soil breathing. This action of the wind is made very evident, too, oftentimes by the rise and fall of a carpet on the floor, and by the buckling of a tin roof as stronger and feebler gusts of wind pass in turn over the house.

In certain sections of the country which are underlaid by extensive beds of coarse gravel, the wells sunk into these beds are often subject to strong draughts, which alternately pass into and out of them. In Sauk County, Wisconsin, there is a district of this sort where the air

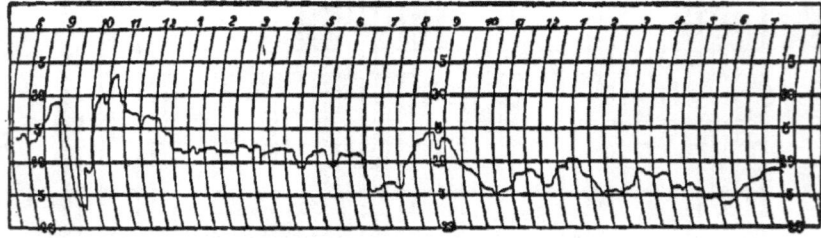

Fig. 37.—Showing the oscillations of water in a well during 24 hours, most of which were coincident with corresponding changes in the barometer.

passes into the wells at times of high barometric pressure and in such large volumes in cold winter weather as to freeze and burst the suction and discharge pipes of pumps at depths greater than 70 feet; then when a low pressure passes over the well, the outgoing current is so strong that even a hat may be lifted by it, and in winter the snow about the well is melted away.

There are many prairies in various parts of the world which are underlaid with layers of coarse gravel, beginning at three to ten feet below the surface and often extending to depths of many feet. The soils of such districts must be subject to a peculiarly strong ventilation or breathing, which may have not a little to do with the wonderful productiveness usually characteristic of such lands. It will be evident, too, from what has been said, that sandy and otherwise light and open soils, especially when they are deep and the distance to the water table is large, will be subject to peculiarly thorough ventilation or breathing, while all close-textured soils, like the heavy and stiff clays, will naturally be aërated less perfectly.

WAYS OF INFLUENCING SOIL VENTILATION.

This leads us to consider some of the means which are available to control soil breathing. It may be said at first that all methods of tillage which tend to develop large, open, non-capillary spaces in the soil must greatly facilitate the change of air, not only in that layer of soil so loosened, but at the same time in that which comes in contact with it below; and in the more thorough soil breathing which deep tillage and the plowing in of coarse manure insures to the stiff clay soils, we find a large part of the good results which follow these treatments.

Then when heavy lands are underdrained, these soils are so much better and more deeply aërated that we must look to this more complete ventilation for the chief advantage which this type of land improvement assures. The aëration of the soil which is rendered possible by thorough underdraining has rarely been sufficiently emphasized, neither has the method by which it is brought about been fully stated.

When the level of the ground water is permanently lowered 2 or 3 feet, as is done in underdraining, the roots of plants penetrate more deeply into the field, and, as they die and decay, leave a system of passage-ways leading toward the surface, into and out of which the soil air finds a more ready ingress and egress. Earthworms, ants, and other burrowing animals also penetrate the ground more deeply, and thus form open ventilating flues of much larger magnitude than those left by the roots of plants.

Then, again, as the under clays dry out, they shrink upon themselves and develop unnumbered fissures, through which the air more freely moves with every change of pressure and temperature, and in which the roots of plants place themselves, the better to profit by renovated air. With the deeper and more thorough penetration of the soil air, carrying with it the carbonic acid developed near the surface, it acts through the soil water upon the lime, producing the bicarbonate, which in its turn tends to flocculate the finer silt particles of the clay, causing them to congregate into larger compound grains and thus render the soil more open in its texture and hence better drained and better aërated, as well as more easily and thoroughly occupied by the roots of plants.

But all of these changes which have been referred to as resulting from thorough drainage are only means for widening and rendering the direct effect of underdrains more prompt in their influence and far-reaching in the renewal of soil air which they secure. In an underdrained field where lines of tile are laid 3 to 4 feet deep and 50 to 100 feet apart, there is provided a ventilation system which operates in an effective manner to hasten the change of the soil air. It must be evident that when the diurnal changes in the temperature of the confined soil air and the changes due to alterations of barometric pressure are sufficiently pronounced to affect the rate of flow of water from tile drains, as has been demonstrated in another place, there must also be draining out of a tile system, at times of rising soil temperature and of falling barometer, considerable quantities of soil air. This must be so because, as the soil air expands and pushes upon the soil water, forcing it out, some portion of the air itself must escape into the drains through their upper sides. Then when the temperature falls and the barometer rises, air will be forced into the soil to take the place of that which had been lost, not simply from the surface of the ground, but simultaneously along the whole extent of the system of underdrains. It is important to recognize in this connection, that the air which underdrains admit to the soil from beneath is in a large measure atmospheric air, containing the normal amount of oxygen, and since under these conditions the soil can breathe free air both from above and below the zone occupied by the roots of plants, it is plain that the soil of a tile-drained field must be much more thoroughly ventilated than that of another entirely similar field having its water table at the same distance below the surface and equally open in texture, but without tile drains.

A word should be said regarding the aërating power of clover as compared with the similar action of other plants. The roots of the red clover being larger than those of cereals, and in part fleshy, tend to separate the soil particles farther, leaving more effective air passages and drainage channels when they decay than do the roots of wheat, oats, barley, or rye. But there is another way in which all free-nitrogen-fixing plants help to aërate the soil during their periods of growth. As these withdraw from the soil air the free nitrogen it contains and fix it in the solid or liquid form, a reduction of air pressure is produced and a fresh supply of air must be crowded in to supply this deficiency. In so far as other plants work in a similar manner to withdraw oxygen and fix it in solid or liquid form, without putting in its place an equal volume of some other gas, a similar tendency must result, but in a less pronounced degree than in the leguminous plants which remove both oxygen and free nitrogen.

Just as the withdrawal of water through underdrains forces air to follow it more deeply into the vacated spaces, so must the osmotic pressure of all roots, by forcing water out of the soil to be evaporated from the leaves, tend also to bring fresh supplies of atmospheric air into the soil, equal in volume to the water withdrawn.

It may well be that some soils are so open that they are too thoroughly ventilated, just as they are too thoroughly drained. In such cases the decaying organic matter is converted into nitrates more rapidly than they can be used by the crop, which results in impoverishing the soil of its nitrogen store.

Soils of this type are helped by shallow tillage and a thorough firming of all soil except so much as shall be needful for a mulch. Keeping such soils thoroughly moist diminishes their porosity, as does farmyard manure, which tends to clog the pores.

The loss of the fine dust particles from light soils so liable to occur in the spring when the winds are strong is a great injury to such lands, both on account of the point here under consideration and because the water-holding power is greatly decreased thereby, as has been pointed out.

HYGROSCOPIC MOISTURE.

The moisture which exists in the air in the form of vapor is to a greater or less extent absorbed by soils as air enters or comes in contact with the ground during the process of soil breathing. The water so taken up by soils or other objects is spoken of as hygroscopic moisture. Usually the finest grained soils, in the air-dry condition, contain more moisture than the coarser grained samples do, but the amounts so retained do not appear to hold any discovered numerical relation to the amount of such surface. It is usually more the lower the soil temperature, and decreases in quantity as the temperature rises during the day. Our knowledge of the laws governing the hygroscopic moisture of soils, and of the importance of this moisture to plant life, is as yet too indefinite to permit any very trustworthy statements regarding it to be made. There are eminent writers who hold that its importance, especially in dry climates, is very great, while others feel that vegetation can derive but little profit from it.

CHAPTER X.

FARM DRAINAGE.

ENOUGH has been said regarding the influence of underdrainage on the important matters of soil temperature, soil ventilation, and the conservation of soil moisture, to show how important this method of improvement may be on certain lands. But as land drainage is certain to play a much larger part in the future of agriculture than it has in the past, it is important that something more should be said regarding it.

We have in the United States, according to Professor Shaler's estimate, more than 100,000 square miles of swamp lands lying east of the 100th meridian; and these, when reclaimed by drainage, are destined to become the most productive lands we have. They will be so, not only because of the large stores of humic nitrogen which, under proper methods of tillage, may be turned into forms available to higher plants, but also because they lie in humid climates, and are so related to the water table that only surplus rains need be lost. Indeed, not only can there be no percolation of soil water from these lands below the reach of root action; but in very many, if not in the majority of instances, they are perennially supplied with water through the underflow from the surrounding higher land, and are thus naturally sub-irrigated.

That our swamp lands are destined to become rich

agricultural fields must be evident when we consider that during the 44 years following 1833 the country of Holland added to itself lands exceeding in square miles the combined area of Rhode Island and Delaware, by systems of dikes and methods of drainage so difficult as removing the surplus water over the sea barriers with pumps. And so fertile are these reclaimed swamp lands that nearly 20 years ago there lived on the 12,731 square miles, a population averaging, for each square mile, 302 people, 20 horses, 118 cows, 70 sheep, 12 goats, and 27 hogs; and, side by side with this population, there were raised 2907 square miles of field crops, exclusive of pastures and hay, and the average yearly product of these fields between 1871 and 1875 is placed at $45.36 per acre.

There are lands other than the swamp areas referred to above which may be much improved by draining; and it may be said, in general, regarding these that all lands will be made more productive by draining where standing water in the ground may be found at seeding time not more than 4 feet below the surface. All very flat areas of fine-textured soil underlaid at four to six feet or less with a stratum of highly impervious clay or rock, and ponds and slews, as well as springy hillsides, are likely to profit much from a thorough system of drainage.

SOME LARGE DRAINAGE SYSTEMS.

As an instance of the pains which in some parts of this country are now being taken to improve lands by underdraining, reference may be made to work which is being done in the state of Illinois, where, in many parts, the fields for long distances are so flat that it is difficult to obtain adequate fall or to provide suitable outlets for the drains when laid. Here, in many instances, the citizens combine their energies and resources and dig broad, open ditches, sometimes many miles in length and deep enough to provide suitable outlets for under-

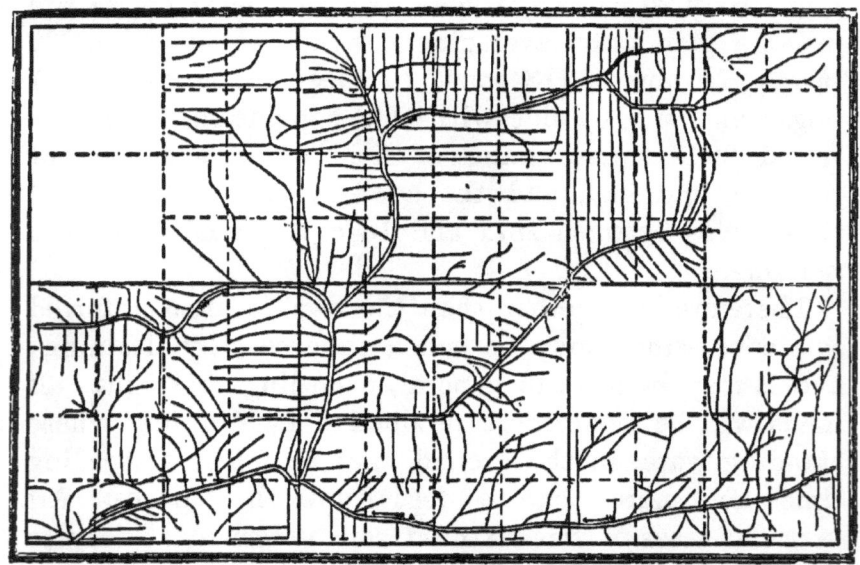

FIG. 38.—Showing plan of the drainage of lands of the Illinois Agricultural Company, Rontoul, Illinois. After Professor J. O. Baker. The smallest squares represent 40 acres; double lines show open ditches; single lines are tile drains.

drains. One such drainage system is found in Mason and Tazewell counties. It was begun in 1883 and completed in 1886, and has a main ditch 17.5 miles long with a width of 30 to 60 feet at the top and a depth of 8 to 11 feet; while, leading into this main channel, there are five laterals averaging 30 feet wide at the top, and from

7 to 9 feet deep, the whole system embracing 70 miles of open ditch.

A clearer idea of the character and magnitude of some of these drainage systems may be gained from Fig. 38, where the double lines indicate open ditches, and the single ones tile drains, many of which it was found necessary to lay very nearly level. This system was begun in 1881 and completed in 1884, and its effect upon the total yield of grains of all kinds is stated by Professor Baker as follows:—

Total yield of grain in 1881, 26,057 bushels.
" " " " " 1882, 58,647 "
" " " " " 1883, 92,360 "
" " " " " 1884, 113,660 "
" " " " " 1885, 122,160 * "
" " " " " 1886, 202,000 "

It is plain from these figures that there has been a marked influence following the underdraining in this case; but it should not be understood that the yields per acre increased to the extent indicated by the figures, for, before the land was drained, there was much of the area too wet to be cropped at all, and hence the larger totals with successive years are due in part to an increase in the acreage.

It has been pointed out that no lands will produce other than swamp vegetation, unless they have first been more or less perfectly drained, because, until this is done, those biologic processes in the soil necessary to cultivated crops cannot be carried forward. They cannot for several reasons: the temperature is too low; there is inadequate soil ventilation; and there is an

* 400 acres of maize destroyed by a water spout.

insufficient amount of room in which the roots can develop and perform their functions.

There are two other ways in which imperfect drainage works disadvantageously; first, by preventing early seeding and thus shortening the growing season and the amount of available water and other forms of plant food derived from the soil; and second, besides increasing the labor of tillage, it shortens the time in which the work can be performed.

DEPTH OF UNDERDRAINS AND DISTANCE BETWEEN THEM.

The depth to which the water should be lowered by drainage, below the surface of the ground, need seldom exceed 4 feet for ordinary farm crops, while under some conditions the lowering of the water table may be less. We have many cases on springy hillsides and on flat areas between rises of ground, where the water is maintained within 4 feet of the surface for only a short period in the spring. In instances like these, where the water table falls normally 5 to 7 feet below the surface as the season advances, it is only necessary to insure a sufficient drying of the surface 18 inches in which the plants may begin their growth. When this is done, the gradual fall of the water table as the crop grows allows the subsoil to become sufficiently dry and open through the natural process of drainage and evaporation from the surface. Where such lands are deeply tile drained, there is liable to be a needless waste of much water.

In sandy soils, too, and others which are naturally leachy and open, it is not as important to draw the ground water down as low as when the soil is more close and impervious in texture. But in all cases where the water table usually remains as near the surface as 4 feet during the whole summer, the drainage system should be planned to draw the water down at once to from 3.5 to 4 feet below the surface, and hold it there.

There are times when the ground has become very dry, and especially when the soil is a stiff clay and has checked to a large extent as the result of drying, when heavy rains come they percolate so rapidly into the system of tiles that a large share of the water is lost before

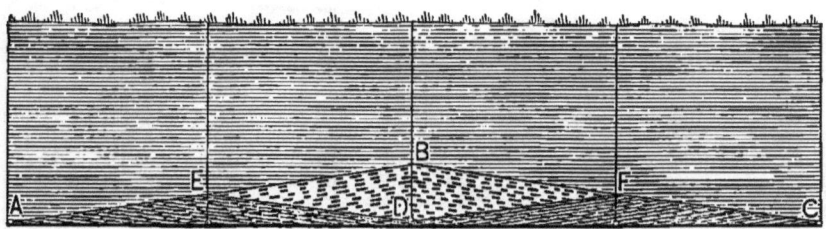

FIG. 39. — Showing how the distance between drains affects the depth of drainage. With drains at *A* and *C* the water will be highest at *B*; but with drains at *A*, *D*, and *C* the surface of the ground water will be more nearly the line *AEDFC*.

it has time to be absorbed by the soil. In such cases, although no provision is usually made for it, the writer feels that, were tile drains provided with a few valves at different places in the system, which might be closed at such times and thus retain the bulk of the water for a day or more, until sufficient time has elapsed for the water to be taken up by the capillary pores of the soil, no inconsiderable advantage would be derived from it. The writer has seen, on three different occasions when a heavy rain has followed a dry period, that on tile-drained ground a very large part of the water was lost through the drains when the soil was much in need of more than had fallen.

The distance between underdrains will vary with the closeness of the soil texture and the depth at which they are laid. Since the water table rises as the distance from the outlet increases, it is plain that midway between lines of tile the surface of the ground water must approach nearer to the top of the ground, and the nearer, the farther the lines of tile are apart, as illustrated in Fig. 39.

The actual contour of the water table in an under-

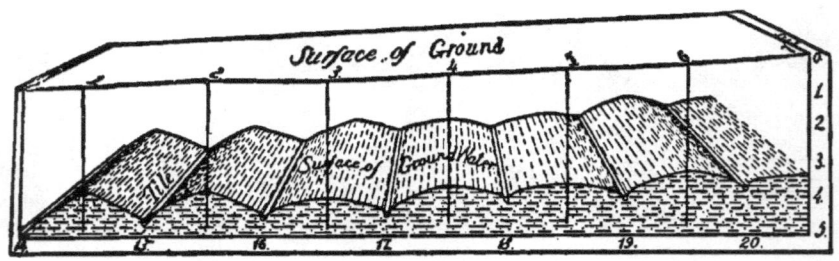

Fig. 40. — Showing the observed surface of the ground water in a tile-drained field 48 hours after a rainfall of .87 inches.

drained field, where the lines of tile are placed at distances of 33 feet and 4 feet below the surface of the ground, is shown in Fig. 40, which gives the contours as they existed 48 hours after a rainfall of .87 inches. In this case the height of the water midway between the lines of tile varied from 4 inches to 12 inches above the tops of the tile, and the mean rate of rise was 1 foot in every 25 feet; that is to say, in soils of this character, when the drains are laid 50 feet apart, the water may stand in the ground midway between the lines 1 foot nearer the surface than the tiles themselves 48 hours after such a rain ; and if 100 feet apart, then 2 feet nearer the surface of the ground. In well 29, Fig. 21, situated 150 feet from the lake, and hence from the drainage outlet, the water stood on June 27, 1892, 7.214 feet above the lake level, thus showing a rise of 1 foot in every 24.4 feet; but later in the season, when the ground had become drier, the level had fallen until the rise was 1 foot in 35.86. As this slope occurred in September, when percolation from the surface had long since ceased, it may be regarded for such soil as a minimum gradient. Taking the rise as 1 foot in 36 feet, lines of tile 72 feet apart and 3 feet deep would allow the water table to rise to within 2 feet of the surface along a line midway between the drains.

SUB-IRRIGATED LANDS.

Just how quickly the water table may be drawn down after a rain by a system of tile drains is shown in Fig. 41, where the broken lines represent the surface of the ground water on May 12, 13, and 16, the latter date being 5 days after a rainfall of .87 inches. It will be seen that the changes in level of the water table were very far from being uniform in different parts of the field, the fall being fastest under the highest ground, where it passed entirely below the upper three lines of tile. In this case, as indicated by the arrows, the lower portion of the field is subject to sub-irrigation, the water under the higher ground tending, through its greater hydrostatic pressure, to move toward and up into the soil of the lower field.

Sub-irrigated lands of this character occur in many places and under conditions where the geological structure is much as represented in Fig. 42. In these cases the surrounding high lands are more or less open in texture, so that the rains percolate into them readily, but drain away slowly, making the adjacent flat lands more or less springy or marshy, and only fit for tillage after they have been underdrained. Such lands, however, once they are underdrained, become very valuable on account of their abundant water supply. Nor is this type of land at all uncommon in the northern part of the United States. Indeed, the glacial hills referred to in an earlier chapter are impounding reservoirs of great extent and capacity, into which the rains sink immediately, and are

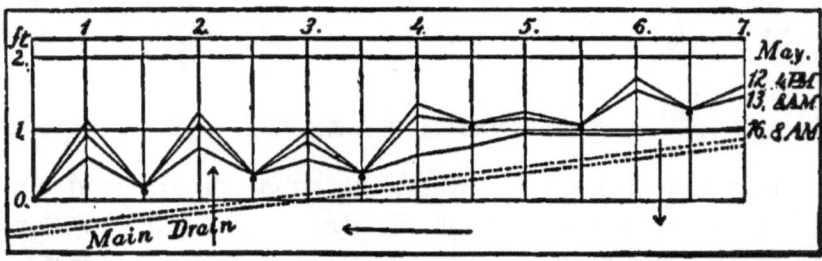

FIG. 41.—Showing the rate of change in the level of the ground water after a rainfall of .87 inches.

there stored, under conditions of least possible loss by evaporation, to be given out gradually in restricted but innumerable areas. Heavy rains which in countries of different structure are lost to agriculture in disastrous floods are here safely and economically stored; and it is to this stored water escaping slowly again from the ground, more than to direct rainfall and flat topography, that we owe the existence of our innumerable small lakes, and the many areas of swamp and lowland pastures so characteristic of glaciated regions. These many naturally sub-irrigated tracts are especially promising for market gardening and other forms of intensive farming.

There is another phase of this question to which attention should be called. There are many and extended tracts of country underlaid by artesian waters, where water can be had at the surface whenever the overlying impervious strata are penetrated. Now, in view of the fact that the best of Portland cements are not wholly impervious to water, even under moderate pressures, it may also be true that in many parts of artesian districts there is a slow, upward penetration of the deeper waters into the field soils, which possibly contributes not a little to the natural productiveness of such lands.

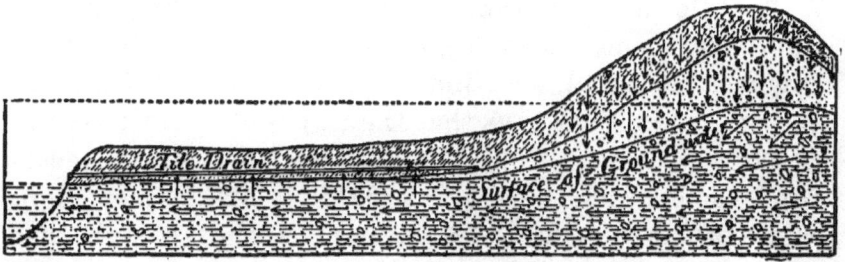

FIG. 42.—Showing the geologic structure favorable to natural sub-irrigation.

SURFACE DRAINAGE.

Where extensive flat fields of very fine-textured, impervious soils occur, it is often desirable, and even necessary, to adopt some form of surface drainage. There are some clayey soils in their virgin state which are so impervious to water that it would be useless to lay drains deeply in them, even were the lines near together, because the water would percolate into them too slowly to render them efficient. In such cases surface drainage must be resorted to. This is done by plowing the fields in narrow lands, leaving dead furrows extending in the direction of the natural slope, and from 20 to 60 feet apart. Where the lay of the land makes it necessary to do so, the dead furrows may be connected by cross furrows, in order to lead the water away through some low place.

It is often desirable, even where underdrains are used, to have some surface drains to carry away surplus water in times of freshets. Such ditches are best made wide with very sloping sides, so as to be grassed over and to permit a team and wagon to be driven across them at any place.

The Celtic land beds of olden times represented an effort to secure deep drainage with surface or open ditches. Marshall, writing of this method of land improvement in 1796, says that the ridges of that time were 8 yards wide and from 2 to 2.5 feet high, but he measured some which were 15 yards wide and 4 feet high, while others were 20 to 25 yards wide, and so high that a horseman riding in one ditch could hardly see his companion riding in the other. This method of drainage must be very wasteful of land, and is only to be resorted to when, for any reason, covered drains cannot be made.

There are many depressions or low places which, on account of being surrounded on all sides by higher land, cannot be made dry either by surface or underdraining. It is frequently true of such cases that the water is held by a pan of clay which has been formed from washings from the higher ground, while under this pan the soil is open and capable of readily carrying the water away. When this is true, and the area which drains into the basin is not too large, it may be drained by boring or digging through the clay in one or more places, making exits for the water to escape by percolation downward.

These drainage ways may be kept open, or they may be filled in with coarse stone, sand, and gravel.

When the clay is too deep to permit of draining in this way, the result may be accomplished, if the amount of water to be removed is not too large, by sinking a well or reservoir of considerable size at the lowest place, into which the drains lead from various directions. This water may then be lifted by wind or other power, and used to irrigate grass or other fields bordering the wet

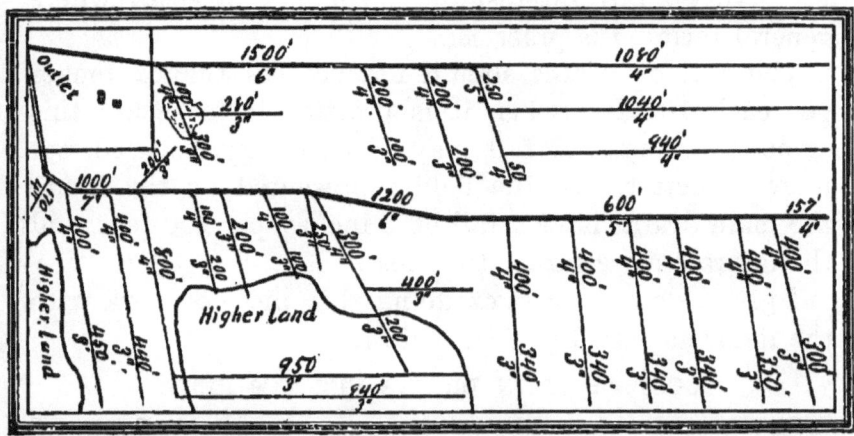

FIG. 43. — Showing the drainage system of 80 acres in northern Illinois. After C. G. Elliott, C.E. Double lines represent mains; single lines are laterals. Numbers give length of drains and size of tile.

area, thus getting rid of the water by evaporation and making it do service at the same time.

COST AND ARRANGEMENT OF UNDERDRAINS.

We cannot in this place discuss the practical details of underdraining, but as an illustration of the method of distributing and joining main drains with laterals, we have selected an actual piece of work done where a farm of 80 acres in northern Illinois has been drained under the supervision of Mr. C. G. Elliott, C.E. The land he describes as a rich black loam, approaching muck in the ponds and flats, underlaid with a yellow clay subsoil at a depth of 2.5 feet from the surface. In draining this piece the object has been to fit it for growing corn, grass, and grain in all seasons, and it will be seen (Fig. 43) that the laterals are, where nearest, about 150 feet apart; but it should be understood that the aim has not been to provide perfect drainage, but rather as good as would pay a fair interest on the money invested where general farming is practiced.

The fall of drains should not be less than 2 inches per each 100 feet when it is practicable to secure this amount. A smaller fall may, if necessary, be used, but more is better. In the field represented by the figure the main drains have a fall of 2 inches per 100 feet, and the laterals more rather than less.

I give below the cost of doing this piece of work with the items as stated by Mr. Elliott.

COST OF MAIN DRAINS PER 1000 FEET.

No. of Feet.	Size.	Depth.	Tile.	Digging, Laying, and Filling.	Total.	Cost per Rod.
1000	7 in.	5 ft.	$60.00	$37.20	$97.20	$1.60
2700	6 in.	5 ft.	40.00	36.60	206.82	1.26
850	5 in.	4 ft.	30.00	24.20	46.07	.89
COST OF LATERAL DRAINS.						
8280	4 in.	3.5 ft.	20.00	20.00	331.20	.06
7030	3 in.	3 ft.	13.20	20.00	233.40	.55
			Total		$914.69	

The total cost of $914.69 makes an average per acre of $11.43.

It will seldom be best to use tile smaller than 3 inches, and in laying them great care should be exercised to place them on a true grade, because if the lines of tile have high and low places in them, whose differences exceed the diameter of the tile, silt will tend to collect in the low places and sooner or later close them up.

The outlet of a drain should be carefully made and end above water as represented in Fig. 44. If the outlet is below standing water, the silt which is brought down tends to close up the main, and thus render the whole system useless. So, too, when a lateral is led into

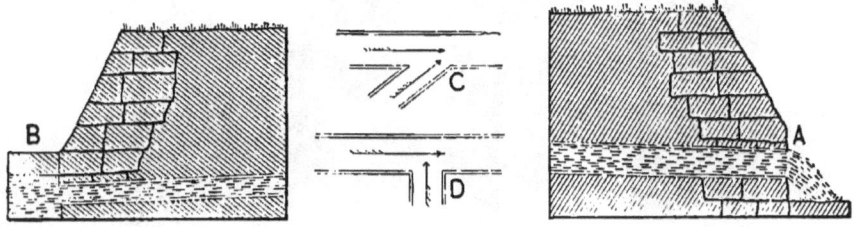

FIG. 44.—Showing proper and improper outlets of drains. *A*, proper outlet; *B*, improper outlet; *C*, proper junction of lateral with main; *D*, improper junction.

a main the union should be made at an angle as at *C* rather than as at *D*.

Care must always be exercised to remove water-loving trees from near lines of underdrains, lest their roots penetrate the joints and there branch into a vast network, entirely filling the tile, when, by retaining the silt brought by the water, they will sooner or later completely close up the tile. Fig. 45 represents a mass of roots thus formed in a tile drain. These roots are those of the European larch, and they penetrated a main drain at a depth of 5 feet below the surface, the trees standing at a distance of 15 feet from the line. As will be seen, the roots, after entering the main, branched into a great system of fibres, which effectually closed a 6-inch drain, making it necessary to take up the line of tile in that vicinity.

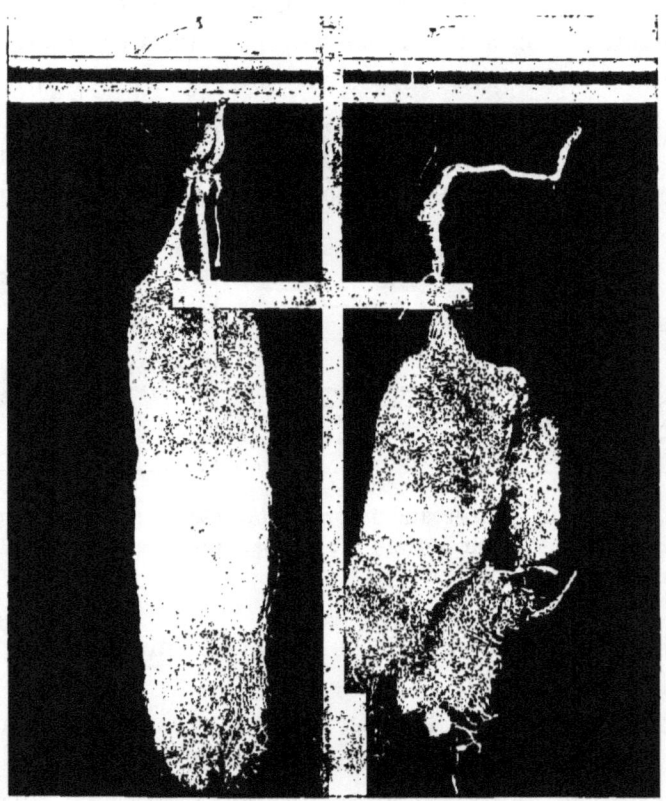

Fig. 45.—Showing the roots of European larch removed from a 6-inch tile drain, which they had effectually clogged.

CHAPTER XI.

IRRIGATION.

REFERENCE was made in the last chapter to the large acreage of lands now lying idle and practically worthless, so far as the needs of civilization are concerned, which have yet to be won to remunerative agriculture through a judicious application of methods of drainage; and to still other lands which may be made more productive than they now are through the same means.

It is not the purpose here to speak of methods of irrigation in detail, nor of irrigation as applied to arid and semiarid regions, because there its value is appreciated and needs no enforcement; but rather to consider in a general way the great promise it has for lands receiving moderate amounts of rain during the growing season. This is done because we are fully persuaded that the maximum limit of productiveness of lands in humid regions can only be attained through a suitable combination of both drainage and irrigation. Further than this, there are many hundred square miles of light lands lying east of the Mississippi, which are now almost as unproductive as the undrained swamps are, but which irrigation, properly handled, would cause to bloom into the richest of gardens. The rainfall of the eastern and central United States is so capricious, the amount of water needed for large yields so great, and the difficulties in the way of making our soils retain enough of the water which falls to meet the demands are so many, that it must be plain to every practical man and student of agriculture, who has devoted much thought to the subject, that the time must come when the waters now running to the sea, with their tons of unused fertility, will be turned to use in irrigating many of the fields through which they flow.

IRRIGATION IN HUMID CLIMATES.

The advantage of irrigating lands in humid climates has been abundantly proved by long trial in many parts of the world, and it is the object here to call attention to the facts and to point out a very important and remunerative channel in which American capital should seek investment. If it will pay to irrigate in arid climates, where all the needed water must be supplied under conditions of cost far exceeding what will be needed in humid districts, there must certainly be many lands near large markets and close to water which can be irrigated with great profit.

In illustration of what sewage irrigation is doing in Scotland, the statements of Storer, made after visiting the irrigated meadows near Edinburgh, may be cited. He says: —

"In 1877 there were 400 acres of these 'forced meadows' near Edinburgh, and they are said to increase gradually. The Craigentinny meadows, just now mentioned, were about 200 acres in extent, and they had been irrigated for thirty years and more. They were laid down at first to Italian ray grass and a mixture of other grass seeds, but the artificial grasses disappeared long ago, couch grass and various natural grasses having taken their place. The grass is sold green to cow keepers, and yields from $80 to $150 per acre. One year the price reached $220 per acre. They get five cuts between the 1st of April and the end of October. This farm of 200 acres turns in to its owner every year from $15,000 to $20,000 at the least calculation, and his running expenses consist in the wages of two men, who keep the ditches in order. The sewage he gets free. The yield of grass is estimated at from 50 to 70 tons per acre. The total produce of the sewage irrigation at Edinburgh amounts to at least $30,000 per annum, taking one year with another. The grass goes to some 2000 cows, and the milkmen all acknowledge that they cannot get any milk-producing food to compare with it for the same amount of money, notwithstanding the seemingly high price that is paid for the grass per acre. Of course the dung from these cows goes to fertilize other farm land."

The lands from which these large returns are being realized are described as having been a worthless, sandy waste previous to their improvement by sewage irrigation.

Referring to results obtained on the Myremill farm of 508 acres near Maybole, in Ayrshire, Scotland, John Wilson quotes at length from the "Minutes of Information" issued by the General Board of Health, to the effect that some 400 imperial acres were laid down to iron pipes placed from 1.5 feet to 2 feet below the surface, and provided with hydrants at intervals, to which hose are attached for the distribution of the water containing the fertilizers. In this case the water used in irrigating the farm was lifted 70 feet with pumps driven by a 12-horse-power engine and stored in large reservoirs, the cost of which, including pipes and hose, is placed at $7676. Counting 7.5 per cent on this sum, for interest and wear and tear, and a working expense in distributing the water of $786.50 per annum, the total annual cost of irrigation equals $1362, or $2.68 per acre. The land laid down to grass is said to have produced 70 tons of green weight per acre, so that at this rate the 70 acres in Italian ray grass gave a gross crop of 4900 tons, and the market value of this one crop of meadow grass exceeded, by a large sum, the first cost of the irrigating plant. It is said that this same land before being treated in the manner described would barely pasture 5 sheep or 1 bullock to the acre, but under the irrigation system it was easy to keep 20 sheep or 5 bullocks to the acre by hurdling and moving from place to place.

Fabulous as these results appear, fourteen other cases are cited by the same author, and they all have the same import. To illustrate: In one case a weighed crop of 10 tons per acre of ray grass was taken, and this is said to be the lightest of four cuttings in one season. In another, where 12 stacks per annum were formerly taken, now 80 are obtained. And again where the yield was worth $1 per acre, it is now $58.

It should be understood that all of the cases here referred to are looked upon as methods of manuring the land, that is, fertilizers are added to the water and with it distributed over the fields; but there can be no question but that the larger yields secured are due much more to the water than to the fertilizers added to it, or more exactly to a proper combination of both rather than to either one of them alone.

In proof of the possibilities of obtaining very heavy yields of dry matter per acre, it may be stated that the writer grew, under field conditions in 1894, by irrigating, 14.5 tons of dry matter to the acre on $\frac{1}{18}$ of an acre of ground, when the crop was a variety of flint corn. The ground upon which this maize was grown was a clover sod, well fertilized with a dressing of farmyard manure, and the same ground with the same variety of maize gave less than 4 tons to the acre where only the natural rainfall was available.

The irrigation of grass lands has been widely practiced in Europe and for a long time. For pasture land and meadows the system is in use in Germany, Switzerland, France, Spain, and in many provinces of Italy. We are told that in 1856 the old kingdom of Sardinia had 600,000 acres of land under irrigation; in Lombardy there were 1,100,000 acres and 300,000 in France. Marsh states that two canals in Lombardy, which now irrigate some 250,000 acres, were dug in the twelfth century. Palestine and Persia are noted for the extensive and highly perfected systems of irrigation which brought wealth to those countries in the days of Babylonian greatness, but which are now in utter ruin. So, too, in Ceylon the native rulers in ancient times covered the whole face of their island with a network of irrigation reservoirs, through which Ceylon became the great granary for southern Asia, but through the devastation of wars they have long since passed into ruins, and it is said that what were once highly productive irrigated fields are now swampy wastes or dense forests.

Even in Mexico and Peru, the Spaniards brought destruction to elaborate systems of irrigation, as they had done before to those in their native land which had been built in the sixth century by the hands of the thrifty Moors.

H. M. Wilson, in speaking of the extent of irrigation, places the acreage in various countries as follows: Total area irrigated in India, 25,000,000 acres; in Egypt, 6,000,000 acres; and in Italy, 3,700,000 acres. In Spain there are 500,000 acres, in France 400,000, and in the United States 4,000,000, acres of irrigated land. In addition to this, he states that there are some millions more of acres cultivated by the aid of irrigation in China, Japan, Australia, Algeria, and South America.

COST OF IRRIGATING AND AMOUNT OF WATER USED.

The amount of water used in irrigating is very large, as indeed it must be when large fields are to be dealt with. In Italy often as much as 4 inches of water on the level are applied to the meadows at a single wetting, and amounts equivalent are applied at intervals of ten days or two weeks, varying of course with the weather; but it usually happens, in these cases, that a considerable portion of the water passes beyond the field to which it is directly applied and is utilized on lower lying areas. Smith concludes from statistics in India that there 1 cubic foot of water per second is sufficient for 180 acres of land where the water is used the whole year through, and this is equal to 48 inches on the level. Five estimates of the amounts of water required in French and Italian experience for water meadows, where the period of consumption is about half the year, gives an average of 86 acres of land watered by a continuous flow of 1 cubic foot per second, and this amounts to covering the ground with a depth of water equal to about 50 inches, the water being used from the middle of March to the middle of September. This is a very large amount of water when considered in connection with the natural rainfall, amounting to from 20 to 40 inches, but it must be observed that three and sometimes four crops of grass are cut from the land each season, besides using it for pasture in the fall.

In obtaining water for irrigation purposes, there are many methods which may be used and are to-day in use in various parts of the world. The most natural method, where the topography of the country will permit of it, is to lead the water out from some point up stream into irrigating canals, which convey the water to flat lands farther down. Or various forms of lifting wheels are placed in the stream, and considerable quantities of water are raised by them and discharged into side channels. Hydraulic rams are used and a great variety of pumps driven by steam or other power. Using a No. 4 rotary pump driven by an 8-horse portable farm engine, the writer has drawn water through 110 feet of 6-inch suction pipe, raising the water to a height of 26 feet, at the rate of 80,320 cubic feet per ton of soft coal, which is equivalent to $22\frac{1}{8}$ inches of water per acre, or over 7 acres covered to a depth of 3 inches. But this amount is much less than could have been moved with the same fuel had the pump been provided with a larger discharge, and could the water have been used as rapidly as pumped, so as to have made frequent stops unnecessary. Windmills may be used to advantage in many cases where the lift is small, where the area to be watered is not large, and particularly where reservoirs of considerable capacity may be used for storage.

While it is true that much profit may be realized through irrigation in humid climates on a small scale, yet the largest returns can be secured only when a considerable capital is involved, but this is no argument against the development of the system, provided the income is sufficient to keep the capital profitably employed.

The cost of irrigation west of the 100th meridian in the United States, as given in the United States census for 1890, averages $8.15 per acre for the construction of canals to bring water to the land, while the average value of the water per acre, when once there, is placed as high as $26, but the average annual cost of water per acre is $.99. The value of the lands before irrigation improvements were added is placed at from $2.50 to $5.00 per acre, but after the water was added to it the same lands are valued at $83.28 per acre.

To show what has been done on barren sands in Belgium, it may be stated that 5636 acres of sand dunes have been reclaimed by irrigation, under government control, and that these lands yield 6615 pounds of hay per acre, with an aftermath valued at $2.00 per acre, which, with the hay, gives an income placed at $29.93 per acre.

CHAPTER XII.

THE PHYSICAL EFFECTS OF TILLAGE AND FERTILIZERS.

The great importance of good tilth has always been appreciated by thoroughgoing practical men, and experience has abundantly taught that the stiffer and more resistant the soil, the greater should be the care and attention given to the field to bring it into perfect tilth before receiving the seeds or plants which it is expected to bring to maturity.

But Nature neither plows, harrows, nor hoes her fields, and yet where water is abundant and the temperature right, the grasses thrive, the flowers bloom and fruit, and tree and shrub vie with one another for the occupancy of the whole surface of the earth even to vertical, rocky cliffs and steep mountain sides. Why, then, should tilth be so important a matter in successful farming?

When we reflect upon Nature's methods, we see plainly that they are quite different from those adopted by thrifty husbandry. In the first place, by Nature's method's, not one seed in many hundreds ever germinates and comes to maturity, but in farming no such chances can be taken. Conditions must be favorable not only for every seed to germinate, but also for each plant to come to full maturrity and bear fruit. In the second place, by Nature's methods, almost all fields bear a mixed vegetation, and her rotations are maintained by slipping in a different plant where the accidents of life, or death from age, have left vacant places; but in agriculture certain crops must occupy the field for the season to the exclusion of all others, and in these differences in aims and methods we find the chief needs for, and the great importance of, good tilth and thorough tillage.

THE IMPORTANCE OF GOOD TILTH.

From what has been said regarding the thorough occupancy of the soil by the roots of our cultivated crops, penetrating as they do three and four feet into the stiff clay subsoils, which have never been stirred, it may seem that a mellow seed bed should have no significance so far as the growth and spread of roots are concerned. Practical experience, however, proves beyond question that a mellow seed bed is important, and a little reflection will make it clear why it must be so.

In the first place, when seeds are beginning to grow, or when sets are transplanted into a new soil, there is yet no organic connection between the plant and the soil upon which it must depend for the supply of indispensable moisture; both the plant in the seed, and the set, have limited stores of nourishment which can be devoted to the development of a root system, placing that in vital connection with the soil grains, and these stores must be used up quickly, once the draft upon them has been begun; for, if they are not, pathogenic changes set in, which result in their destruction and the death of the seed or the set. Now, if the seed is placed under conditions where it finds many obstacles in the way of a free and symmetrical development of the first roots, it loses 'time and food is wasted in getting so connected with the soil as to be able to feed itself. A mellow seed bed, with its many well-aërated pores, allows the roots to grow unhindered in any and every direction, and to place their absorbing surfaces in vital touch with the soil grains and soil moisture. In this way the nourishment in the seed produces the maximum root surface in the shortest time, which is an evident and great advantage.

In the second place, our methods of tillage tend inevitably to so alter the texture of the surface soils, especially if they are heavy, as to make the spread of young roots through them more difficult, and hence thorough stirring for tilth becomes more important than it was in the virgin state. The frequent stirring tends to break down the compound grain structure, so that the action of rains, and of stirring when too wet, causes the soil grains to run together into masses of so close a texture that the young roots find difficulty in making their way among them, and are insufficiently supplied with air even if they succeed in doing so. It is this physical change forced upon heavy clay soils which makes it so essential that they be laid down frequently to grass and given time for bringing together again into compound grains the minute particles which frequent tillage tends to separate, and which the rains cause to run together into masses of close texture.

The perfect tilth and freedom from clods so characteristic of virgin soils, is always more or less completely restored whenever soils have been laid down to grass for a sufficient length of time. After they are covered with sod, the puddling action of rains is prevented, and, as the roots grow and decay, the soil particles are wedged apart in some places and crowded together in others. Then the solvent action of carbon dioxide in the soil water results, through the deposit of the dissolved lime or other materials, in cementing together the grains and in restoring the more open and mellow texture characteristic of virgin soils. It is plain from these facts that the laying down of land to grass at frequent intervals is beneficial in other ways than that of increasing the stores of certain kinds of plant food, and rotation of crops is seen to have a significance in the good influence it has upon soil texture.

MANAGEMENT OF SOIL TO SECURE GOOD TILTH.

When stirring soil to improve its texture, the amount of moisture present at the time plays a very important part in the final result. If water enough is present to nearly or quite fill all the capillary spaces, leaving no free water surface upon which surface tension can come into play, then the individual soil particles move over one another with the least resistance, and a little pressure or stirring at such a time causes them to slip into all large empty spaces and to assume the most compact arrangement possible, but one very unfriendly to the normal life processes going on in the soil. When the amount of water becomes less, however, so that free water surfaces are formed in all the larger non-capillary pores, then surface tension comes into action, tending to bind the particles of the compound soil grains and smaller shrinkage lumps together, and plowing or otherwise stirring the soil in this stage causes it to crumble and assume that open texture so much sought for by those who appreciate the importance of good tilth.

It sometimes happens, however, to clayey soils, which are naked, that they become excessively wet through drenching rains, and if, under these conditions, hot, drying winds follow, so that water is evaporated from the surface rapidly, an internal stress or strain is set up, which tends to force the soil grains into the places vacated by the water. Now if the drying is very rapid at the time when the soil grains float together most easily, that is, when the 5 or 6 inches below the surface the soil is nearly full of water, then certain spots which, for any reason, chance to be more resistant than others, come to be centres toward which the surrounding soil grains begin to move under the pull of surface tension, like the walls of a soap bubble, and cracks are formed bounding large blocks of soil, which are forming into clods. The width of the cracks shows how much the internal pore space of the blocks has been reduced, and how much the texture of the soil is being injured by the destruction of needed passage ways.

But if the drying after such wet times is slow, giving time for the water to drain away as well as to evaporate, then the soil moves upon itself with more difficulty, and the effect of surface tension is to cause shrinkage about very many rather than a few centres, and the formation of large clods is averted. This being true, it is plain that the formation of large clods under the former conditions may be prevented by stirring the surface just as soon as the ground is dry enough to permit this to be done. Stirring the surface, as with a harrow or cultivator, at such times operates in two ways to prevent the shrinkage into large blocks. It lessens the loss of water at the surface, and thus gives more time for drainage before much drawing together has taken place, and also by cutting or scratching the surfaces of large blocks which have begun to form, they tend to divide it into smaller sections on account of weakening due to the surface cutting.

So, too, if the ground is plowed at such times, as soon as the soil is dry enough not to puddle, then, as the furrow slice is bent and doubled upon itself, it tends to divide into innumerable layers, which, in shearing over one another, break and pulverize any clods which have commenced to form. This shearing or pulverizing action of the plow will be seen at once, if the reader will close this book and bend its leaves abruptly upon themselves. As this is done, it will be observed that each leaf slips past or upon the other. Just so it is with the plow, and the steeper the mold board, that is, the more abruptly the furrow slice is bent, the greater is the pulverizing effect; and it follows from this, that soils naturally hard should, as a rule, be plowed with a steeper plow than is necessary or desirable for the lighter and naturally mellow lands.

If the leaves of this book were all glued rigidly together, so that one could not slip upon the other when bent, then no shearing would take place; but under sufficient force, the leaves would break. So, too, with the furrow slice; if the soil has become too dry before plowing, the shearing or pulverizing effect of the mold board is prevented, and the furrow simply breaks into larger or smaller lumps, instead of small crumbs; and not only is a poorer quality of work done, but more energy must be expended to do it. It is evident, therefore, that those who have stiff clay soils to work need to exercise great judgment regarding the condition the soil is in when it is stirred, and these remarks apply to harrowing and cultivation as well as to plowing.

In the case of corn ground, if a heavy rain has fallen

upon clay land after the crop is planted, great effort should be made to stir the surface of that soil just as soon as the team can walk upon it without sinking into it more than 1 to 1.5 inches, and the heavier the soil, the more important it is that just the right moment be seized upon. If a light harrow is used in such cases, so that a very shallow surface layer is moved, it is surprising to see how soon after a rain such land may be stirred, and how helpful this slight stirring is in preserving the open, mellow texture as well as needed moisture. It is at such times as these that very shallow cultivation for heavy soils is specially to be recommended, but to be followed by deeper stirring as soon as the ground is drier.

We have already pointed out how important good tilth is in securing the right temperature, adequate soil ventilation, and less loss of water by evaporation.

From what has been said here, and in other places, it follows that subsoiling for the sake of improving texture will be desirable only in special cases. Well-drained subsoils, which have been long under the influence of vegetation and the action of burrowing animals, like the earthworms and ants, appear to have become sufficiently porous to meet the demands of most crops in humid regions, so that deeper tillage than that needful to set a crop fairly upon its feet is unnecessary. The depth of surface tillage, for the sake of texture, should vary somewhat with the crop, soil, and season. Where crops with fleshy roots are to be grown in heavy soils, it becomes specially important to secure an open texture in order that the rapidly expanding tubers may find it not too difficult to crowd the soil aside and make room for themselves, and it is the difficulty in maintaining a sufficiently yielding soil on the heavy clay, which makes hill or ridge culture preferable many times to the level tillage more generally used on the loose, friable lands.

In the semiarid regions, where irrigation is not practiced, and where the soil is fertile to considerable depths, deep tillage, preparatory to seeding, and deep planting are sometimes desirable to produce a texture so open that the scanty rains may enter the soil deeply and at once, in order that it shall not be lost by surface evaporation, and that capillarity shall not return too much water to the surface, as has been referred to. In such cases the deep, open texture and methods of listing allow seeds and roots adequate ventilation with a likelihood of more water.

The texture of clayey soils, when fall plowed and exposed to winter freezing, becomes sensibly altered and usually for the better, a more crumbly and friable character being the result. As the soil water in wet clays freezes, it is withdrawn from the interspaces to a considerable extent, and built into crystals of varying sizes, among the soil grains, in such a manner as to fissure and crumble the clay, and if in the spring drenching rains do not undo, by puddling, the action of the ice crystals, the soil is left more open.

It is not improbable, however, that the action of carbonic acid, in its tendency to flocculate the colloidal clay, may play a very important, if not the chief, part in transforming the cold, obstinate disposition of these soils into a more friendly nature during the winter weathering.

BURNING AND PARING.

It has been a common practice in some countries to improve the texture of heavy clay soils by burning, and the allied operation of paring and burning had, at one time, an even more extended use.

Properly burnt clay loses its plastic quality and falls easily into a light, friable powder, and those clays which contain silicate of potash, with some carbonate of lime, are most improved by the process.

It is essential in the burning of clay that it be not allowed to pass much above a dull red heat; for otherwise it becomes hard and brick-like and loses its chemical vigor to a large extent. When so treated, clay loses its power of again becoming adhesive when wet, while it apparently imparts the same quality to a considerable volume of that which has not been burned, thus giving to the whole a much more open texture and less adhesive quality.

It has long been known to the arts that raw clay is not easily acted upon, even by strong acids, and so in the manufacture of chemicals the inert clay is exposed for some time to a dull red heat, and in this process it comes to be readily acted upon by sulfuric acid in the operation of alum making. Whether the burning has any other effect than to destroy the colloidal clay, and thus render the mass easily penetrated by the acids, and hence more readily acted upon, does not appear to be well understood. But it is true that the clay becomes much more porous, more easily worked and attacked by chemical agents; and these are valuable qualities in any agricultural soil. In burning clay, it is usually mixed with brushwood or peat or sometimes with soft coal. To better control the operation and hold the fire in check, long, narrow trenches are sometimes dug and then filled in with alternate layers of brushwood and clay and, after being fired, more clay and fuel added as the burning proceeds.

The allied process of paring and burning seems now to have largely gone out of use, though in parts of Europe it was quite general on certain lands. It consisted in shaving off 1 to 3 inches of the surface soil with its sod or stubble and gathering it together into heaps to be burned, when dry enough, with the aid of the roots, stubble, and other organic matter it may have contained. The process was, of course, attended with considerable losses of nitrogen, and hence was wasteful besides being laborious and costly, but it did produce marked beneficial effects on many soils.

It seems that liming, and the thorough underdraining of such lands, which have come into vogue, have now rendered these methods of tillage nearly or quite obsolete.

TEXTURE OF SOILS INFLUENCED BY FERTILIZERS.

There appears to be a large number of substances, and among them many of the chemical fertilizers, which have an appreciable influence in altering the texture of the soil, making it more or less open and friable. Among those which have the power of flocculating colloidal clay, lime has been most generally recognized, and it appears that this may be applied to the soil either as the oxide, hydrate, or carbonate, with the same ultimate effect, though perhaps with varying rates of action. We have already referred to Hilgard's and Schlösing's observations regarding this action of lime.

More recently R. Sachsse and A. Becker have experimented with lime-water, gypsum, sulfate of magnesia, sulfate of ammonia, chlorides of potassium, sodium, and ammonium, and carbonic, hydrochloric, nitric, sulfuric, and silicic acids, as regards their power to bring about a compounding of minute silt particles into clusters constituting grains of larger size. Working with a nearly pure kaolin, consisting of particles so small that only .5 per cent of them were as coarse as .0004 of an inch, they found one part of lime would flocculate 20.7 parts of kaolin which had remained suspended in 27.777 parts of water. A similar effect was observed upon a heavy clay loam, but not upon a humus alluvial loam. On examining the sizes of the particles, after flocculation had taken place, it was found that in the case of the kaolin and clay loam, the diameters of very many of them had materially increased, but that in the case of the alluvial soil no appreciable change had occurred.

Studying the effects of lime on the percolation of water through kaolin and the clay loam, the persons referred to above found that while no water would pass through the untreated clay loam, yet when 61.2 cc. of water were poured over the tube treated with lime, it reached the bottom of the column in 1 hour and 10 minutes, and 22.5 cc. drained through. In the case of the kaolin, of the 40 cc. added, 19 drained from the limed tube, while only a few drops percolated from the one untreated, and in that the kaolin swelled so much as to finally burst the tube.

Carbonic acid flocculated the kaolin readily, and so did the hydrochloric, nitric, and sulfuric acids. While land plaster and the sulfates of magnesia and of ammonia were found to have the same power to some extent, their influence was not marked. Common salt and the chlorides of potash and ammonia also flocculated the clay and the kaolin, but the nitrate of soda had the opposite effect.

If it is true that nitrate of soda has a tendency to

destroy the flocculation of clay soils and render them closer in texture, it would follow that, when supplies of nitrogen need to be added to these soils, they should be given in some other form than the Chile saltpetre.

Either directly or indirectly, fertilizers exert an influence upon the relation of water to the soil, as, indeed, has been implied in what has been said regarding their power to make the texture of the soil finer or coarser. If, by making the soil coarser grained, water percolates through it more readily, and its water-holding power is decreased, it should be expected that the capillary power will also be influenced; and observations have already been cited, showing that potassium nitrate increases the amount of water lifted by capillarity through a column of sand, while solutions of lime, land plaster, and salt decrease the amount, when compared with distilled water. Now, these effects may be brought about through changes in the texture of the soil or through a strengthening or weakening of the surface tension. In the cases we have cited, on a former page, however, only the effect of the potassium nitrate is readily explained by either of these modes of action, taking Whitney's measurements of surface tension as a basis, because all of these salts are given as increasing it, while only one of them increased the rate of capillary rise through the sand; which cannot be assumed to have been materially altered by flocculation.

When fertilizers are applied, the soil may react upon them either chemically or physically, and in such a manner as often to wholly prevent or greatly diminish their loss in drainage waters at times when percolation is taking place. Many observers have noted that, on filtering salt water through sand, the first which comes through is nearly free from salt, but when any soil becomes saturated, so to speak, with any particular fertilizer, then all excess may be leached away, and this fact has an important bearing on the application of fertilizers to lands, and shows that lighter dressings, often repeated, are likely to be less wasteful than heavier ones applied at longer intervals. This statement has even greater importance when applied to the organic fertilizers, like farmyard manure; for the richer a soil is in organic matter, the more rapidly will fermentation be set up in it under good tillage, and under these conditions nitrification is liable to be more rapid than is required to meet the demands of the crop on the ground. But if this is true, the soil becomes charged so that when percolation takes place a large loss of nitrogen may occur. It is plain, therefore, that it is better to spread the manure over a larger area in the right amount than to concentrate it in heavy dressings on small areas.

INFLUENCE OF FARMYARD MANURE ON SOIL MOISTURE.

Farmyard manure has a marked effect upon the amount and disposition of soil moisture in cultivated fields. When coarse manure is plowed under, its first effect is to act as a mulch to the unstirred soil, by breaking the capillary connection between it and the surface layer. The tendency, therefore, is to cause the surface soil to become drier than it would otherwise have become in the same time, and frequently to an injurious extent, especially in times of spring droughts, before seeds have germinated or young plants have developed a root system reaching into the deeper soil. On such occasions as these, the heavy roller is of service in making a better capillary connection with the unstirred subsoil.

Farmyard manure has a general tendency to leave the upper three feet of soil more moist than it would be without it, and the drier the season and the more thorough the manuring, the more marked is its influence. In experiments to measure this influence under field conditions, the writer has found, as a mean of 3 years' work, that for manured fallow ground the surface foot contained 18.75 tons more water per acre than adjacent and similar but unmanured land did, while the second foot contained 9.28 tons and the third 6.38 tons more water, making a total difference in favor of the manured ground amounting to 34.41 tons per acre. The largest observed difference was 72.04 tons in the dry season of 1891. Early in the spring, on ground manured the year before and fallow, there was an observed difference amounting to 31.58 tons per acre.

It is a fact long ago observed that increasing the organic matter in the soil increases its water-holding power, and this being true, it was to be expected that the surface foot would be more moist as a consequence of manuring, but by referring to the table on the next page, it will be seen that the influence of the manure is felt below the level to which it is applied.

A part of the observed difference in the water content of manured and unmanured soils is to be explained by the fact that the rate of surface evaporation is diminished by the dressing of dung. In the case of two cylinders 42 inches deep and holding nearly 600 pounds of soil, one manured and the other not, the writer found that the unmanured cylinder lost, in 105 days, at the rate of 108.5 tons more water per acre, a difference of about

TABLE SHOWING EFFECT OF FARMYARD MANURE ON THE WATER CONTENT OF THE SOIL AT DIFFERENT DEPTHS BELOW THE SURFACE.

DATE.	SURFACE FOOT.		2D FOOT.		3D FOOT.		4TH FOOT.		5TH FOOT.		6TH FOOT.	
	MANURE.	NO MANURE.	MANURE.	NO MANURE.	MANURE.	NO MANURE.	MANURE.	NO MANURE.	MANURE.	NO MANURE.	MANURE.	NO MANURE.
	Per cent.		Per cent.		Per cent.		Per cent.		Per cent.		Per cent.	
July 22, 1891	17.37	16.41	19.37	17.63	17.03	16.36	12.00	12.78	14.69	15.40	18.01	20.06
Sept. 12, 1891	13.75	12.76	16.41	14.89	15.67	14.49	12.96	11.59	11.49	10.96	15.61	15.99
Apr. 11, 1892	23.56	22.86	20.66	20.55	18.26	17.50	16.50	16.53	16.88	17.53	19.72	19.41
July 13, 1892	25.89	24.36	24.01	23.87	23.72	23.29	23.74	23.32				
Aug. 30, 1892	25.64	24.76	24.42	24.34	24.21	23.74	21.69	21.72				
Aug. 12, 1893	19.07	17.91	17.85	17.78	16.87	17.88	17.45	17.99				
Sept. 9, 1893	13.55	12.50	16.13	16.22	16.90	16.94	16.71	17.30				
Mean . . .	19.88	18.79	19.79	19.33	18.88	18.60	17.29	17.32	14.35	14.63	16.98	17.13
Difference .	1.09		.46		.28		.03		.28		.15	
Tons per acre	18.75		9.28		6.38		.69		6.75		3.63	

1 ton per day. In another case wetting the surface of sand with water leached from manure reduced the rate of evaporation from the surface from 64.98 pounds per unit area to 32.72 pounds in the same time, under otherwise identical conditions.

But the table shows that farmyard manure exerts an influence different from simply decreasing the surface evaporation; for, while the upper three feet of the dunged land is more moist than that not so treated, the reverse is true when the next three feet are compared. Indeed, results analogous to wetting the surface soil and to firming it are observed. It appears as though the deeper soil water had been brought nearer to the surface, where it may become available for crop production, and in the writer's opinion this really does take place.

On manured ground which is producing a crop, while a much larger yield of dry matter is produced per acre, the soil at the end of the growing season is found only a little drier, from which it follows, either that less water is required to produce a pound of dry matter in the rich soil, or else that the manure in some way makes more water available, and when the matter comes to be thoroughly understood, it is not improbable that both propositions will be found true.

PHYSICAL AND CHEMICAL EFFECTS OF FALLOWING.

That form of tillage known as fallowing exerts marked physical and chemical effects upon a soil, not felt, at least with like intensity, on lands heavily cropped. One of the most marked effects produced by fallowing is that exerted upon the water content of the soil. Not only is the fallow ground more moist during the fallowing period, as indeed should be expected, but the influence is felt the following spring, and even at the end of the harvest after the crop has been removed from the ground.

The writer found, for example, in the spring succeeding a summer fallow, after all of the fall, winter, and spring rains, that the land which had been fallowed contained, in its upper four feet, 9.35 pounds to the square foot, or 203 tons per acre, more water than did that which had been cropped the season before. Nor was this all; for at the end of the growing season and after large crops of oats and barley had been harvested from the land, there was still a difference in the water content of the upper four feet amounting to 8.21 pounds on both the oat and barley ground, or 179 tons per acre. That the differences here recorded were not due to inherent differences in the soil, is proved by the water content of the same lands taken at three different times before the fallowing experiments began. The mean of these three determinations showed a difference of only .7 pounds in favor of the piece which was subsequently fallowed, as against 9.35 pounds in the spring and 8.21 pounds after the harvest succeeding the fallowing. It is plain, therefore, that summer fallowing a piece of land exerts a strong influence over its relation to soil moisture, and one which is not outgrown for more than a year.

Common experience, too, shows that the productiveness of the same lands is much increased, nor is it strange that this should be so. For when we call to mind the large stores of comparatively inert nitrogen which all good soils have been shown to possess, and then reflect upon the conditions which are most favorable to the nitrification of this humus, it must be very evident that in the warmer, more moist, and frequently stirred and aërated fallow field there must be a much larger conversion of inert nitrogen into active forms than could possibly take place in the soil so thoroughly occupied by the roots of a crop; that neither sufficient moisture nor other forms of food are available for the full and vigorous development of the microscopic soil life so indispensable to the processes here in question. When a soil is sapped to dryness and depleted of its soluble phosphates, lime and potash, the action of microscopic life in it must come nearly to a standstill and the development of fertility nearly cease.

The Lois-Weedon system of tillage, which the Rev.

Mr. Smith practiced in Northamptonshire for many years, was simply a judicious application of summer fallowing and a revival, in modern form, of the doctrine taught by Jethro Tull, between 1680 and 1740, that thorough and proper tillage might take the place of manure.

Mr. Smith laid off his fields in lands 5 feet wide, and on these in successive years, by alternation, he grew wheat continuously, raising the yield per acre from 16 bushels to 34, as an average of many years. In this rotation the fallow strips were frequently and deeply stirred, which favored the complete utilization of the native fertility of his soil and that left in the ground by the roots and stubble of the crop harvested.

In adopting this system, Mr. Smith left a considerable portion of his land where nitrates could be developed in it under the best of conditions, and at the same time where the plants along the margins of the tilled lands could reach their roots out into the enriched soil and avail themselves of both its native moisture and its nitrates as they formed.

The practice now in vogue in parts of Europe of growing grain in drills far enough apart to permit of cultivating the soil between the drills with horse cultivators, is carrying the idea of Tull and of Smith to the extreme limit, and is, of course, placing the growing of these crops on the same basis as that which we follow with so much success with corn, potatoes, and the like.

The special advantage which these methods of summer fallowing, for such they in reality are, have over the old method of broad fields, is that the fertility developed may be utilized by the crop the same season, and thus the risk of losing a part of it through leaching by the winter and spring rains may be largely reduced.

In very wet climates or more especially in those which have heavy rainfall outside the growing season, so that excessive percolation and loss of plant food through drainage is large, summer fallowing in broad fields cannot be recommended. But in dry countries, where the loss of plant food through drainage channels is small, broad-field summer fallowing may in some cases prove decidedly advantageous, because, with the deficient rainfall, there may not be moisture enough to mature a paying crop and at the same time to develop a sufficient store of plant food from the native fertility of the soil to meet the demands of the next season. At all events, the arguments urged against fallowing in countries like England do not apply to the semiarid districts of the world with equal force.

INDEX.

A.

Abbot, Col., cut-offs in Mississippi River, 46.
Actiniæ, symbiotic life of, 129.
Aëration of soil, influenced by drainage, 76.
Aëration of soils, 239.
Air and soil, 239.
Air in soil retards percolation, 173.
Algæ, symbiotic life of, 129.
Aluminum, 79.
Amazon, volume of discharge, 24.
Ammonia, in air, 12; brought down by rain and snow, 119; absorbed by soil, 122; in air on Montsouris, 123; decomposed by micro-organisms, 131.
Ammonium carbonate, absorbed by plants from the air, 122.
Arendt, distribution of silica in plants, 97.
Argon, amount of, in the atmosphere, 11.
Arid soils, texture, 29, 31; effect of lime in, 30, 94; amount of lime in, 30, 94; chemical composition of, 84, 85, 86, 87; compared with humid soils, 92.
Ash ingredients, of plants, 101; relation between demand and supply, 113.
Asphalt, origin of, 20.
Atmosphere, its work, 9; pressure of, 9; relation of, to eating, drinking, and breathing, 10; depth of, 11; composition of, 11, 23, 123; retains solar heat, 12; selective power of different constituents, 14; a distributing agent, 15; currents in, 23.
Atwater, Prof., analysis of night-soil, 133.

B.

Bacteria, nitrifying, 125; observations of Frank, Schlösing, Jr., Laurent, 129; producing ammonia, 131.
Baker, J. O., quoted, 255.
Barley, roots of, 208, 212.
Barley, water used by, 155.
Barometric pressure, oscillation due to changes in, 179.
Beans, fixation of free nitrogen by, 124; water used by, 155.
Becker, A., quoted, 285.
Bisulfid of carbon, absorption of solar energy by, 8.
Bitumen, origin of, 20.
Blue grass, roots of, 212.
Borax, 79.
Boron, 79.
Buckwheat, water used by, 156.
Burning land, 283.

C.

Calcium, 80.
Capillarity, 135, 138; measured in

horse power, 142; the work of sunshine in, 146.
Capillary movement, 142; of soil water, 173; rate of, 174; in dry soil, 176; decreased by plowing, 177; influenced by dissolved salts, 177.
Capillary tubes, height of water in, 140.
Carbon, in soil, 77.
Carbon dioxide, in air, 11; corrosive power of, 19; flocculation by, 31, 76, 78.
Catch crops, in relation to moisture, 190.
Celtic land beds, 263.
Chemical analyses of soils, 84, 85, 86, 87; interpretation of, 89.
Chile saltpetre, 81.
Chlorine, 78, functions, 99; in rain water, 121.
Chlorophyll, importance of iron, 98.
Clay, in soil, 28; flocculated by lime, 30; colloid, 30; flocculated by carbon dioxide, 31; influence of drainage on texture of, 76; alumina in, 79; influence on percolation, 172.
Clayey soil, compared with sandy soil, 90; yields water to plants with difficulty, 161; contamination of well water in, 169.
Clover, ash ingredients, 102; increases soil nitrogen, 118, 124; water used by, 155.
Clover, roots of, 208, 212.
Coal, formed by life, 20.
Cold waves, relation to storms, 14.
Colloids, movements of, 153.
Color of soil, and warmth, 230.
Color waves, 7.
Conservation of energy, 21.
Conservation of moisture, 184.
Corn, amount of water used by, 155.
Corn, roots of, 208, 209.
Craigentinny meadows, 269.

Crocker, amount of nitrogen in cultivated soil, 107.
Cultivating to conserve moisture, 192.

D.

Darwin, G. H., tidal friction, 17.
Darwin, Charles, action of earthworms, 62.
Denitrification, 130, 132, 133.
De Vries, observations on osmotic pressure, 148.
Diffusion, 147.
Drainage, 253.
Drainage, influence of, on soil texture, 76.
Drains, movements of water in, due to soil air, 179.
Dry-earth closets, 133.
Dust, retards radiation, 14.

E.

Early seeding, 190.
Earth, cooled by water, 16.
Earthworms, 61.
Eel grass, 67.
Electricity, production of nitrous and nitric acids by, 123.
Elliott, C. G., quoted, 264, 265.
Erosion, by rains, 50; by glaciers, 55, 58.
Ether waves, velocity of, 6; complex character of, 7; absorbed by bisulfid of carbon and alum water, 8; absorbed by water, 8.
Evaporation and soil temperature, 225.
Evaporation, caused by sunshine, 4, 145; nature of, 143; unnecessary amount from plants, 156; retarded by deposit of salts on soil, 175.
Evolution, organic, 31; relation of soil to, 31; of true roots, 35; of woody fibre, 36; of forest trees, 36.

F.

Fallowing, effects of, 291.
Fall plowing, 187.
Farm drainage, 253.
Farming, need for improved methods, 3.
Farmyard manure and soil moisture, 288.
Fertilizers and soil texture, 285.
Fertilizers, effects of, 276.
Fertilizers, experiments of Sir J. B. Lawes, 105.
Flocculation, by lime, 30; by carbon dioxide, 76.
Fluorine, 79.
Forced meadows, 269.
Frank, nitrifying bacteria, 129.
Free nitrogen-fixing germs, 124.

G.

Gases, precipitation of minerals by, 20; origin of natural, 20.
Glaciers, origin of, 54; in soil formation, 55; extent of action, 56; mode of action, 58; moraines of, 60; magnitude of work, 61.
Glauconite, 80.
Granite, water absorbed by, 39.
Greensand, 80.
Gypsum, 78.

H.

Hauksbee, early studies of capillarity, 136.
Harrowing, to conserve moisture, 192.
Heat, absorption and radiation of, 3; causing air currents, 4; part played in solution, 144.
Hellriegel, fixation of free nitrogen, 124; fixation of nitrogen by lupines, 126; water used by plants, 135; best amount of soil moisture, 161.

Hiawatha, quoted, 211.
Hilgard, E. W., effect of lime on clay, 30; lime in arid soils, 30, 94; mechanical analyses of soils by, 70; composition of humus, 96.
Hilgard, quoted, 285.
Horse power, measure of solar energy expressed in, 6.
Humid soils, 64; texture of, 29, 31; formation of, 64; in lakes, 66; by rivers, 65; chemical composition of, 84, 85, 86, 87; compared with arid soil, 92.
Humus, 94; relation to fertility, 94; origin, 95; richer in nitrogen in arid regions, 96; Hilgard and Jaffa on composition, 96.
Hydrogen, 78.
Hygroscopic moisture, 252.

I.

Iceland moss, 34.
Illinois, drainage in, 254.
Iron, 81; importance in plant life, 98.
Irrigation, 268.

J.

Jaffa, composition of humus, 96.

K.

Kainite, 81.
Kaolin, potash in, 81.
Kelvin, Lord, 6.
Kosswitsch, nitrifying bacteria, 129.

L.

Lakes, overgrowing, 66; formation of swamp soils in, 66; oscillations of, 179.
Land plaster, 78; decreases capillary movement, 177.

Langley, Professor, observations on atmospheric absorption, 13.
Lathyrus sylvestris, roots of, 212.
Laurent, nitrifying bacteria, 129.
Lawes, Sir J. B., experiments on the influence of fertilizers, 105; nitrification of soils, 109.
Leguminous plants, fixation of free nitrogen by, 124.
Leonardo da Vinci, phenomena of capillarity, 136.
Level culture and moisture, 202.
Lichens, 34; examples of symbiosis, 128, 130.
Life, relation of water to, 18; its work in rock and soil building, 18, 20; microscopic forms in soil, 19; formation of coal, peat, etc., 20; accumulation of phosphorus by, 79; symbiotic, 128; processes producing nitric acid, 131.
Lime, effect on soil texture, 30; flocculation by, 30; in arid soils, 94; amount in soil, 102; decreases capillary movement, 177.
Limestones, formed by life, 20; composition, 80.
Loess, 69; texture of, 75; chemical composition, 84, 85, 86, 87.
Lois-Weedon system, 292.
Lupines, fixation of free nitrogen by, 124; experiments by Hellriegel, 126.

M.

Madison River, shiftings of its course, 48.
Magnesia, amount in soil, 102.
Magnesium, 80.
Maize, roots of, 208, 209.
Maize, water used by, 24, 155.
Manganese, 81; brown oxide of, 93.
Manitoba, amount of nitrogen in soils, 108.
Manna, 34.

Manure and moisture, 288.
Marshall, quoted, 263.
McGee, W. J., Mississippi bad lands, 50.
Mechanical analysis of soil, 71, 90.
Microscopic life in soil, 19, 20, 37; in fixation of nitrogen, 124; in the production of nitrous and nitric acids, 131; in denitrification, 132; deoxidation by, 132.
Mississippi, bad lands, 50; soils, 70.
Mississippi River, materials borne in solution, 17, 25; volume of discharge, 24; solids moved by, 25, 48; bends in, 46; ox-bones, 46.
Moisture and manure, 288.
Moisture, conservation of, 184.
Moisture, in air retards radiation, 14.
Montsouris, analysis of air on, 123.
Moraines, 60.
Moss, Iceland, 34; Reindeer, 34; Tripe de Roche, 34; sphagnum, 66.
Mother of petre, 241.
Mulches, to conserve moisture, 194.
Müntz, deoxidation in waterlogged soils, 132.
Myremill farm, 270.

N.

Nernst, W., nature of solution, 145.
Night soils, denitrification of, 133.
Nitrates, deoxidation of, in soils, 115; effects of percolation on distribution, 117; destruction of, 131, 132; movement of, in plants, 150.
Nitre farming, 241.
Nitric acid, 79; constituent of the atmosphere, 12; corrosive power of, 19; amount in soil, 114; in rain and snow, 119; produced by electricity, 123; life processes producing, 131.

Nitrification, 130.
Nitrogen, 11, 79; amount in the soil, 107; in soils of Manitoba, 108; importance of, in plant growth, 111; relation between demand and supply, 113; forms of, in the soil, 114; stored as humus, 115; distribution in the soil, 115; sources of, in soil, 117; in soil increased by clover, 118; brought down by rain and snow, 119; fixed by microscopic life, 124; fixation by lupines, 127; Wagner's experiments, 127, 129; loss of, in night soils, 133.
Nitrous acid, produced by electricity, 123; produced by nitrous ferment, 131.
Nitrous ferment, 131.
Nollet, Abbé, early observation on osmosis, 148.

O.

Oak, roots of, 214.
Oats, amount of water used by, 155.
Oats, roots of, 208, 212.
Ocean, mean depth of, 16.
Ochre, red and yellow, 81.
Oil, origin of mineral, 20.
Organic matter, deoxidation of, in soil, 115.
Osmosis, 147; experiments in, 148.
Ostwald, magnitude of surface tension, 138.
Outlets of drains, 266.
Oxygen, 11; in soil, 77.
Ozone, 12; action in producing nitrous and nitric acids, 123.

P.

Paring land, 283.
Peas, fixation of free nitrogen by, 124, 126, 128; amount of water used by, 155.
Pelouze, influence of size of grains on rate of solution, 74.

Percolation, effect on distribution of nitrates, 117; rate of, from sand, 158, 171; of water into wells, 167; direction of, 170; rate of, through different soils, 171; observations by Wollny, 172; retarded by air in soil, 173; effected by barometric changes, 179.
Peroxide of hydrogen, action in the production of nitrous and nitric acids, 123.
Pfeffer, W., observations on osmotic pressure, 148.
Phosphoric acid, amount of, in manure, 74; amount in soil, 102.
Phosphorus, 78; importance in plant life, 110.
Physical effects of tillage, 276.
Plants, relation to soil formation and soil destruction, 19; relation to types of soil, 100; ash ingredients, 101; circulation in, 149; movement of nitrates in, 150; movements of starch, gum, and fats, 151; selective power in, 152; amount of water used by, 155; unnecessary evaporation from, 156.
Plant feeding, process of, 149.
Plant growth, part played by sunshine in, 5; importance of nitrogen, 110; amount of nitrogen demanded, 113.
Plowing decreases the rate of capillary movement, 177.
Plowing to save moisture, 187.
Potash, amount of, in manure, 74; distribution, 80; amount in soil, 102.
Potassium, 80.
Potassium carbonate, effect on capillary movement, 178.
Potassium nitrate increases capillary movement, 177.
Potatoes, amount of water used by, 155.

Precipitation, nature of, 4.
Pressure of atmosphere, 9; measure of, 9; relation of, to breathing, eating, and drinking, 10; variations of, the cause of winds, 10; variations with light, 11.

Q.

Quartz, abundance of, in soil, 77.
Quincke, surface tension, 139.

R.

Radiolaria, symbiotic life of, 129.
Rainfall, insufficiency of, 25, 156.
Rains, action in soil formation, 50; erosion by, in Mississippi, 50; ammonia and nitric acid brought down by, 119; composition of, 120.
Rains and soil warmth, 233.
Rains for irrigating, 274.
Ramsey, Prof., 11.
Read, T. M., rock dissolved by water, 19.
Reighley, Lord, 11.
Richthofen, formation of loess, 69.
Rivers, action in soil formation, 46, 48; in formation of swamp soils, 65; oscillations in, 179.
Rock, thickness formed by water, 17; structure of, in soil formation, 38; fissures in soil formation, 42; fractured by roots, 42.
Rogers, the Messrs., experiments on the rate of solution, 74.
Rolling, and soil temperature, 235.
Rolling, effect on moisture, 200.
Root hairs, part played in solution, 146; osmotic movement in, 150.
Roots, distribution of, 207.
Roots, evolution of, 35; rocks fractured by, 42, 43; extent of contact with soil, 75; development in light soil, 99.

Roots in drains, 266.
Root surface, 207.
Rotation of solid land, 26; of alluvial soils, 48.
Rothamsted, distribution of nitrogen in soils, 115.
Rye, amount of water used by, 155.

S.

Sachs, observations on function of iron, 98.
Sachsse, quoted, 285.
Salt, in rain water, 121; decreases capillary movement, 177.
Sand, water retained by, 157.
Sandy soil, compared with clay soil, 90; size of grains, 100; water capacity influenced by distance to water table, 160; yields water to plants readily, 161.
Schlösing, per cent of clay in soil, 28; flocculation by carbon dioxide, 31; absorption of ammonia by soil, 122; denitrification, 132.
Schlösing, quoted, 285.
Seeding, early, 190.
Seeds, ash ingredients of, 80.
Shaler, formation of salt marshes, 67.
Shaler, quoted, 253.
Shrinking of soils, 249.
Silica, abundance of, in soils, 77; soluble, 84, 93, 102; functions, 97; distribution in oat plant, 97.
Slope of land, and warmth, 228.
Smith, Dr. Angus, composition of rain water, 120; denitrification in sewerage, 132.
Smith, Rev., quoted, 292.
Snow, ammonia and nitric acid brought down by, 119.
Soap bubbles, illustrating surface tension, 136.
Soda, amount in soil, 102.

Index.

Sodium, 78, 81; chloride decreases capillary movement, 177.

Soil, nature of, 27; clay in, 28; comparison of subsoil with, 29; texture of arid and humid, 29; a storehouse of water, 37; microscopic life in, 19, 37; formation by running water, 45, 46; overplacement of, 53; of glacial origin, 55, 56; prairie, relation to glacial action, 60; swamp or humus, 64; formation by winds, 68; loess, 69; texture of, 70; mechanical analyses of, 71; extent of contact of roots with, 75; chemical constituents, 76; sandy, description of, 82; clayey, 83; chemical composition of, 84, 85, 86, 87; kinds, 99; warm and cold, 99; amount of nitrogen in, 107; distribution of nitrogen in, 115; absorbs ammonia from the air, 122; capillary rise of water in, 139, 141; water capacity, 157, 159; rate of capillary rise in, 174, 176.

Soil air, movements of soil water due to, 178; producing diurnal oscillations of soil water, 180.

Soil grains, size of, 72; influence of size on water capacity, 72, 157, 160; smallest not individually invested by water films, 73; influence of size on rate of solution, 72; compound structure of, 75; in varieties of soil, 100; influence of size on percolation, 171.

Soil moisture, part played in soil fertility, 154; best amount of, 161; conservation of, 177.

Soil temperature, 218.

Soil texture, 70; influence of, on water capacity, 72; influence of, on food supply, 74; influenced by drainage, 76; influence of, on capillary movement, 176, 177.

Soil water, movements of, 170, 173; movements due to soil air, 178; diurnal oscillations of, 180.

Solution, 142; Pelouze and Rogers, experiments on rate of, 74; influence of size of soil particles on rate of, 74; nature of, 142; influence of temperature on, 144; part played by root hairs in, 146.

Sphagnum moss, 66.

Springs, flow of, influenced by barometric changes, 180.

Spring plowing, 188.

Starch, movements of, in plants, 151.

Storer, amount of nitrogen in soil, 107; dry-earth closets, 133.

Storer, quoted, 269.

Sub-irrigation, 260; natural, 171.

Subsoil, 29; chemical composition, 84, 85, 86, 87; compared with soil, 92.

Sulfur in soil, 78, 97.

Sulfuric acid, amount in soil, 102; in rain water, 121.

Sun, work done by, 6.

Sunshine and its work, 3; nature of, 3; causing evaporation, 4; stored in the soil, 5; part played in plant growth, 5; velocity of, 6; measured in horse power, 6; absorbed by bisulfid of carbon, 8; snow and ice opaque to dark rays of, 8; solution and movement of plant food, 144; in capillarity, 146.

Surface drainage, 262.

Surface tension, 136; magnitude of, 138; distance over which it acts, 139; part played in solution, 143; modified by dissolved salts, 177.

Swamp soils, 64; formation by rivers, 65; in lakes, 66; by the sea, 66; by eel-grass, 67; chemical composition, 84, 85, 86, 87.

Symbiosis, 128, 134.

T.

Temperature of soils, 218.
Temperature of the earth without an atmosphere, 13; changes of, in soil formation, 39; influence on solution, 144; influence on water capacity, 160; diurnal oscillations of soil water due to changes of, 183.
Texture of soil, 70; influenced by drainage, 76.
Texture of soils influenced by fertilizers, 285.
Tidal friction, 17.
Tillage and moisture, 192.
Tillage, deep, and moisture, 202.
Tillage, physical effects of, 276.
Tilth, importance of, 277; to secure, 279.
Timothy, roots of, 212.
Translocation of moisture, 195.
Tripe de Roche, 34.
Tubercles on roots of leguminous plants, 124, 125.
Tull, Jethro, 293.

U.

Underdraining, 257.
Underdraining and aëration, 248.
Underdraining and temperature, 225.

V.

Van't Hoff, nature of solution, 145.
Velocity of ether waves, 6.
Ventilation of soils, 239.
Voeleker, composition of farmyard manure, 74.

W.

Wagner, P., fixation of nitrogen by peas, 127; importance of nitric nitrogen, 129.
Waring, Col., dry-earth closets, 133.
Warington, nitrogen removed from soil by wheat, 107; amount of nitric acid in soil, 114; production of nitrous acid, 123; production of nitric acid, 123; denitrification in water-logged soils, 132.
Warmth of soils, 220.
Water in soils, 184.
Water, opaque to dark rays, 8; and its work, 16; cooling the earth, 16; retards earth's rotation, 17; dissolving and transporting power, 17, 19; importance in life processes, 18; magnitude of movement of, 24; absorbed by rocks, 39; freezing of, in soil formation, 39; solvent action of, 39; running, in soil formation, 45; amount used by plants, 155; capacity of soils for, 155; retained by sand, 157; capacity of field soils for, 159; yielded to plants by kinds of soil, 161; contamination of, in wells, 167; percolation into wells, 168; rate of capillary rise in soil, 174, 176.
Water capacity, influence of size of soil grains on, 72; of field soils, 159.
Water films, on soil grains, 72.
Water table, 163; influence on water capacity of soils, 160; position influences land values, 162; altitude of, 163; height variable, 163; wells most affected by seasonal changes, 166.
Waves, ether, 6; velocity of, 6; complex character of, 7; kinds of, 7; absorbed by water, 8.
Weight of soils, 185.
Wells, 162, 167; relation to water table, 163; conditions to be observed in digging, 166; most affected by seasonal changes of the water table, 166; depth below

water table influences capacity, 167; contamination of water in, 167; movements of water in, due to soil air, 179.

Wheat, amount of water used by, 155.

Wheat, roots of, 212.

Whitney, mechanical analyses of soils, 91.

Wiley, mechanical analyses of soils, 90.

Wilson, H. M., quoted, 273.

Wilson, John, quoted, 270.

Windmills for irrigating, 274.

Winds and soil moisture, 204.

Winds, systems of, 23; soils formed by, 68; caused by sunshine, 4; maintain uniform composition of air, 23.

Wolff, ash ingredients of plants, 101.

Wollny, observations on percolation, 172.

www.ingramcontent.com/pod-product-compliance
Lightning Source LLC
Chambersburg PA
CBHW030806230426
43667CB00008B/1083